Statistics for Terrified Biologists

Statistics for Terrified Biologists

Helmut F. van Emden
Emeritus Professor of Horticulture
School of Biological Sciences
University of Reading, UK

Blackwell
Publishing

©Helmut F. van Emden 2008

BLACKWELL PUBLISHING
350 Main Street, Malden, MA 02148-5020, USA
9600 Garsington Road, Oxford OX4 2DQ, UK
550 Swanston Street, Carlton, Victoria 3053, Australia

The right of Helmut F. van Emden to be identified as the author of the editorial material in this work has been asserted in accordance with the UK Copyright, Designs, and Patents Act 1988.

All rights reserved. No part of this publication may be reproduced, stored in a retrieval system, or transmitted, in any form or by any means, electronic, mechanical, photocopying, recording or otherwise, except as permitted by the UK Copyright, Designs, and Patents Act 1988, without the prior permission of the publisher.

Designations used by companies to distinguish their products are often claimed as trademarks. All brand names and product names used in this book are trade names, service marks, trademarks, or registered trademarks of their respective owners. The publisher is not associated with any product or vendor mentioned in this book.

This publication is designed to provide accurate and authoritative information in regard to the subject matter covered. It is sold on the understanding that the publisher is not engaged in rendering professional services. If professional advice or other expert assistance is required, the services of a competent professional should be sought.

First published 2008 by Blackwell Publishing Ltd

3 2009

Library of Congress Cataloging-in-Publication Data

Van Emden, Helmut Fritz.
   Statistics for terrified biologists/Helmut F. van Emden.
     p. cm.
   Includes bibliographical references and index.
   ISBN 978-1-4051-4956-3 (pbk.:alk.paper)
 1. Biometry–Textbooks. I. Title.

QH323.5.V33 2008
570.1'5195–dc22

                                        2007038533

A catalogue record for this title is available from the British Library.

Set in 10.5/12.5 pt Photina
by Newgen Imaging Systems (P) Ltd., Chennai, India
Printed and bound in Singapore
by Fabulous Printers Pte Ltd

The publisher's policy is to use permanent paper from mills that operate a sustainable forestry policy, and which has been manufactured from pulp processed using acid-free and elementary chlorine-free practices. Furthermore, the publisher ensures that the text paper and cover board used have met acceptable environmental accreditation standards.

For further information on Blackwell Publishing, visit our website at
www.blackwellpublishing.com

Contents

Preface

I have written/edited several books on my own speciality, agricultural and horticultural entomology, but always at the request of a publisher or colleagues. This book is different in two ways. Firstly, it is a book I have positively wanted to write for many years and secondly, I am stepping out of my "comfort zone" in doing so.

The origins of this book stem from my appointment to the Horticulture Department of Reading University under Professor O. V. S. Heath, FRS. Professor Heath appreciated the importance of statistics and, at a time when there was no University-wide provision of statistics teaching, he taught the final year students himself. Following Heath's retirement, the course was given by his successor, Professor Peter Huxley, and I was asked to run the practical exercises which followed the lectures. You cannot teach what you do not understand yourself, but I tried nonetheless.

Eventually I took over the entire course. By then it was taught in the second year and in the third year the students went on to take a Faculty-wide course. I did not continue the lectures; the whole course was in the laboratory where I led the students (using pocket calculators) through the calculations in stages. The laboratory class environment involved continuous interaction with students in a way totally different from what happens in lectures, and it rapidly became clear to me that many biologists have their neurons wired up in a way that makes the traditional way of teaching statistics rather difficult for them.

What my students needed was confidence – confidence that statistical ideas and methods were not just theory, but actually worked with real biological data and, above all, had some basis in logic! As the years of teaching went on, I began to realize that the students regularly found the same steps a barrier to progress and damage to their confidence. Year after year I tried new ways to help them over these "crisis points"; eventually I succeeded with all of them, I am told.

The efficacy of my unusual teaching aids can actually be quantified. After having taken the Faculty course taught by professional statisticians, my students were formally examined together with cohorts of students from other departments in the Faculty (then of "Agriculture and Food") who had attended the same third year course in Applied Statistics. My students mostly (perhaps all but three per year out of some 20) appeared in the mark list as a distinct block right at the upper end, with percentage marks in the 70s, 80s, and even 90s. Although there may have also been one or two students from other courses with high marks, there then tended to be a gap until marks in the lower 60s appeared and began a continuum down to single figures.

I therefore feel confident that this book will be helpful to biologists with its mnemonics such as SqADS and "you go along the corridor before you go upstairs." Other things previously unheard of are the "lead line" and "supertotals" with their "subscripts" – yet all have been appreciated as most helpful by my students over the years. A riffle through the pages will amaze – where are the equations and algebraic symbols? They have to a large extent been replaced by numbers and words. The biologists I taught – and I don't think they were atypical – could work out what to do with a "45," but rarely what to do with an "x." Also, I have found that there are a number of statistical principles students easily forget, and then inevitably run into trouble with their calculations. These concepts are marked with

the symbol of a small elephant – "Never forget . . . ."

The book limits itself to the traditional foundations of parametric statistics: the t-test, analysis of variance, linear regression, and chi-square. However, the reader is guided as to where there are important extensions of these techniques. Finally the topic of nonparametric tests is introduced, but the calculation procedures are not explained. This is because the principles of the significant tests and therefore the calculations involved are outside the scope of this book; these calculations and the associated tables are readily available on the internet or in larger textbooks. However, the chapter does include one worked example of each test described as well as a check list of nonparametric methods linked to their parametric counterparts.

Many chapters end with an "executive summary" as a quick source for revision, and there are additional exercises to give the practice which is so essential to learning.

In order to minimize algebra, the calculations are explained with numerical examples. These, as well as the "spare-time activity" exercises, have come from many sources, and I regret the origins of many have become lost in the mists of time. Quite a number come from experiments carried out by horticulture students at Reading as part of their second year outdoor practicals, and others have been totally fabricated in order to "work out" well. Others have had numbers or treatments changed better to fit what was needed. I can only apologize to anyone whose data I have used without due acknowledgment; failure to do so is not intentional. But please remember that data have often been fabricated or massaged – therefore do not rely on the results as scientific evidence for what they appear to show!

Today, computer programs take most of the sweat out of statistical procedures, and most biologists have access to professional statisticians. "Why bother to learn basic statistics?" is therefore a perfectly fair question, akin to "Why keep a dog and bark?" The question deserves an answer; to save repetition, my answer can be found towards the end of Chapter 1.

I am immensely grateful to the generations of Reading students who have challenged me to overcome their "hang-ups" and who have therefore contributed substantially to any success this book achieves. Also many postgraduate students as well as experienced visiting overseas scientists have encouraged me to turn my course into book form. My love and special thanks go to my wife Gillian who, with her own experience of biological statistics, has supported and encouraged me in writing this book; it is to her that I owe its imaginative title.

I am also most grateful for the very helpful, constructive and encouraging comments made by Professor Rob Marrs (School of Biological Sciences, University of Liverpool) and Dr Steve Coad (School of Mathematical Sciences, Queen Mary University of London). They both took the trouble to read the entire text at the request of the publisher and to do this they have given up a great deal of their time. Any misinterpretations of their advice or other errors are my fault and certainly not theirs.

Finally, I should like to thank Ward Cooper of Blackwells for having faith in this biologist, who is less terrified of statistics than he once was.

Helmut van Emden
May 2007

1

How to use this book

Chapter features

Introduction

Don't be misled! This book cannot replace effort on your part. All it can aspire to do is to make that effort effective. The detective thriller only succeeds because you have read it too fast and not really concentrated – with **that** approach, you'll find this book just as mysterious.

The text of the chapters

The chapters, particularly 2–8, develop a train of thought essential to the subject of analyzing biological data. You just have to take these chapters in order and quite slowly. There is only one way I know for you to maintain the concentration necessary for comprehension, and that is for you to **make your own summary notes** as you go along.

My Head of Department when I first joined the staff at Reading used to define a university lecture as "a technique for transferring information from a piece of paper in front of the lecturer to a piece of paper in front of the student, without passing through the heads of either." That's why

I stress **making your own summary notes**. You will retain very little by just reading the text; you'll find that after a while you've been thinking about something totally different but seem to have apparently read several pages meanwhile – we've all been there! The message you should take from my Head of Department's quote above is that just repeating in your writing what you are reading is little better than taking no notes at all – the secret is to digest what you have read and reproduce it in your own words and in summary form. Use plenty of headings and subheadings, boxes linked by arrows, cartoon drawings, etc. Another suggestion is to use different color pens for different recurring statistics such as "variance" and "correction factor." In fact, use anything that forces you to convert my text into as different a form as possible from the original; **that** will force you to concentrate, to involve your brain, and to make it clear to you whether or not you have really understood that bit in the book so that it is safe to move on.

The actual process of making the notes is the critical step – you can throw the notes away at a later stage if you wish, though there's no harm in keeping them for a time for revision and reference.

So DON'T MOVE ON until you are ready. You'll only undo the value of previous effort if you persuade yourself that you are ready to move on when in your heart of hearts you know you are fooling yourself!

A key point in the book is the diagram in Fig. 7.5 on page 55. Take real care to lay an especially good foundation up to there. If you **really** feel at home with this diagram, it is a sure sign that you have conquered any hang-ups and are no longer a "terrified biologist."

What should you do if you run into trouble?

The obvious first step is to go back to the point in this book where you last felt confident, and start again from there.

However, it often helps to see how someone else has explained the same topic, so it's a good idea to have a look at the relevant pages of a different statistics text (see Appendix 4 for a few suggestions, though of course there are many other excellent textbooks).

A third possibility is to see if someone can explain things to you face to face. Do you know or have access to someone who might be able to help? If you are at university, it could be a fellow student or even one of the staff. The person who tried to teach statistics to my class at university failed completely as far as I was concerned, but I found he could explain things to me quite brilliantly in a one-to-one situation.

Elephants

At certain points in the text you will find the "sign of the elephant," i.e. .

They say "elephants never forget" and the symbol means just that – NEVER FORGET! I have used it to mark some key statistical concepts which, in my experience, people easily forget and as a result run into trouble later and find it hard to see where they have gone wrong. So, take it from me that it is really well worth making sure these matters are firmly embedded in **your** memory.

The numerical examples in the text

In order to avoid "algebra" as far as possible, I have used actual numbers to illustrate the working of statistical analyses and tests. You probably won't gain a lot by keeping up with me on a hand calculator as I describe the different steps of a calculation, but you should make sure at each step that you understand where each number in a calculation has come from and why it has been included in that way.

When you reach the end of each worked analysis or test, however, you should go back to the original source of the data in the book and try and rework the calculations which follow on a hand calculator. Try to avoid looking up later stages in the calculations unless you are irrevocably stuck, and then use the "executive summary" (if there is one at the end of the chapter) rather than the main text.

Boxes

There will be a lot of individual variation among readers of this book in the knowledge and experience of statistics they have gained in the past, and in their ability to grasp and retain statistical concepts. At certain points, therefore, some will be happy to move on without any further explanation from me or any further repetition of calculation procedures.

For those less happy to take things for granted at such points, I have placed the material and calculations they are likely to find helpful in "boxes" in order not to hold back or irritate the others. Calculations in the boxes may prove particularly helpful if, as suggested above, you are reworking a numerical example from the text and need to refer to a box to find out why you are stuck or perhaps where you went wrong.

Spare-time activities

These are numerical exercises you should be equipped to complete by the time you reach them at the end of several of the chapters.

That is the time to stop and do them. Unlike the within-chapter numerical examples, you should feel quite free to use any material in previous chapters or "executive summaries" to remind you of the procedures involved and guide you through them. Use a **hand calculator** and remember to write down the results of intermediate calculations; this will make it much easier for you to detect where you have gone wrong if your answers do not match the solution to that exercise given in Appendix 3. Do read the beginning of that Appendix early on – it explains that you should not worry or waste time recalculating if your numbers are similar, even if they are not identical. I can assure you, you will recognize – when you compare your figures with the "solution" – if you have followed the statistical steps of the exercise correctly; you will also immediately recognize if you have not.

Doing these exercises conscientiously with a hand calculator, and when you reach them in the book rather than much later, is really important. They are the best thing in the book for impressing the subject into your long-term memory and for giving you confidence that you understand what you are doing.

The authors of most other statistics books recognize this and also include exercises. If you're willing, I would encourage you to gain more confidence and experience by going on to try the methods as described in this book on their exercises.

Executive summaries

Certain chapters end with such a summary, which aims to condense the meat of the chapter into little over a page or so. The summaries provide a reference source for the calculations which appear more scattered in the previous chapter, with hopefully enough explanatory wording to jog your memory about how the calculations were derived. They will, therefore, prove useful when you tackle the "spare-time activities."

Why go to all that bother?

You might ask – why teach how to do statistical analyses on a hand calculator when we can type the data into an Excel spreadsheet or other computer program and get all the calculations done automatically? It might have been useful once, but now . . . ?

Well, I can assure you that you wouldn't ask that question if you had examined as many project reports and theses as I have, and seen the consequences of just "typing data into an Excel spreadsheet or other computer program." No, it does help to avoid trouble if you understand what the computer should be doing.

So why go to all that bother?

- Planning experiments is made much more effective if you understand the advantages and disadvantages of different experimental designs and how they affect the "experimental error" against which we test our differences between treatments. It probably won't mean much to you now, but you really do need to understand how experimental design as well as treatment and replicate numbers impact on the "residual degrees of freedom" and whether you should be looking at one-tailed or two-tailed statistical tables. My advice to my students has always been that, before embarking on an experiment, they should draw up a form on which to enter the results, invent some results, and complete the appropriate analysis on them. It can often cause you to think again.
- Although the computer can carry out your calculations for you, it has the terminal drawback that it will accept the numbers you type in without challenging you as to whether what you are asking it to do with them is sensible. Thus – and again at this stage you'll have to accept my word that these are critical issues – no window will appear on the screen that says: "Whoa – you should be analyzing these numbers nonparametrically," or "No problem. I can do an ordinary factorial analysis of variance, but you seem to have forgotten you actually used a split-plot design," or "These numbers are clearly pairs; why don't you exploit the advantages of pairing in the t-test you've told me to do?" or "I'm surprised you are asking for the statistics for drawing a straight line through the points on this obvious hollow curve." I could go on.
- You will no doubt use computer programs rather than a hand calculator for your statistical calculations in the future. But the printouts from these programs are often not particularly user-friendly. They usually assume some knowledge of the internal structure of the analysis the computer has carried out, and abbreviations identify the numbers printed out. So obviously an understanding of what your computer program is doing and familiarity with statistical terminology can only be of help.

- A really important value you will gain from this book is confidence that statistical methods are not a "black box" somewhere inside a computer, but that you could *in extremis* (and with this book at your side) carry out the analyses and tests on the back of an envelope with a hand calculator. Also, once you have become content that the methods covered in this book are based on concepts you can understand, you will probably be happier using the relevant computer programs.
- More than that, you will probably be happy to expand the methods you use to ones I have not covered, on the basis that they are likely also to be "logical, sensible and understandable routes to passing satisfactory judgments on biological data." Expansions of the methods I have covered (e.g. those mentioned at the end of Chapter 17) will require you to use numbers produced by the calculations I have covered. You should be able confidently to identify which these are.
- You will probably find yourself discussing your proposed experiment and later the appropriate analysis with a professional statistician. It does so help to speak the same language! Additionally, the statistician will be of much more help to you if you are competent to see where the latter has missed a statistical constraint to the advice given arising from biological realities.
- Finally, there is the intellectual satisfaction of mastering a subject which can come hard to biologists. Unfortunately, you won't appreciate it was worth doing until you view the effort from the hindsight of having succeeded. I assure you the reward is real. I can still remember vividly the occasion many years ago when, in the middle of teaching an undergraduate statistics class, I realized how simple the basic idea behind the analysis of variance was, and how this extraordinary simplicity was only obfuscated for a biologist by the short-cut calculation methods used. In other words, I was in a position to write Chapter 10. Later, the gulf between most biologists and trained statisticians was really brought home to me by one of the latter's comments on an early version of this book: "I suggest Chapter 10 should be deleted; it's not the way we do it." I rest my case!

The bibliography

Right at the back of this book is a short list of other statistics books. Very many such books have been written, and I only have personal experience of a small selection. Some of these I have found particularly helpful, either to increase my comprehension of statistics (much needed at times!) or to find details of and recipes for more advanced statistical methods. I must emphasize that I have not even seen a majority of the books that have been published and that the ones that have helped me most may not be the ones that would be of most help to you. Omission of a title from my list implies absolutely no criticism of that book, and – if you see it in the library – do look at it carefully; it might be the best book for you.

2

Introduction

Chapter features

What are statistics?

"Statistics" are summaries or collections of numbers. If you say "the tallest person among my friends is 173 cm tall," that is a statistic based on a scrutiny of lots of different numbers – the different heights of all your friends, but reporting just the largest number.

If you say "the average height of my friends is 158 cm" – then that is another statistic. This time you have again collected the different heights of all your friends, but this time you have used all those figures in arriving at a single summary, the average height.

If you have lots and lots and lots of friends, it may not be practical to measure them all, but you can probably get a good estimate of the average height by measuring not all of them but a large sample, and calculating the average of the sample. Now the average of your sample, particularly of a small sample, may not be identical to the true average of all your friends. This brings us to a key principle of statistics. We are usually trying to evaluate a *parameter* (from the Greek for "beyond measurement") by making an *estimate* from a sample it is practical to measure. So we must always distinguish *parameters* and *estimates*. So in statistics we use the word "*mean*" for the estimate (from a *sample* of numbers) of something we can rarely measure – the parameter we call the "average" (of the entire existing *population* of numbers).

Notation

"Add together all the numbers in the sample, and divide by the number of numbers" – that's how we actually calculate a *mean*, isn't it? So even that very simple statistic takes a lot of words to describe as a procedure. Things can get much more complicated – see Box 2.1.

> **BOX 2.1**
>
> "Multiply every number in the first column by its partner in the second column, and add these products together. Now subtract from this sum the total of the numbers in the first column multiplied by the total of the numbers in the second column, but first divide this product of the totals by the number of pairs of numbers. Now square the answer. Divide this by a divisor obtained as follows: For each column of numbers separately, square and add the numbers and subtract the square of the total of the column after dividing this square by the number of numbers. Then add the results for the two columns together."

We really have to find a "shorthand" way of expressing statistical computations, and this shorthand is called *notation*. The off-putting thing about notation for biologists is that it tends to be algebraic in character. Also there is no universally accepted notation, and the variations between different textbooks are naturally pretty confusing to the beginner!

What is perhaps worse is a purely psychological problem for most biologists – your worry level has perhaps already risen at the very mention of algebra? Confront a biologist with an "x" instead of a number like "57" and there is a tendency to switch off the receptive centers of the brain altogether. Yet most statistical calculations involve nothing more terrifying than addition, subtraction, multiplying, and division – though I must admit you will also have to square numbers and find square roots. All this can now be done with the cheapest hand calculators.

Most of you now own or have access to a computer, where you only have to type the sampled numbers into a spreadsheet or other program and the machine has all the calculations that have to be done already programmed. So do computers remove the need to understand what their programs are doing? I don't think so! I have discussed all this more fully in Chapter 1, but repeat it here in case you have skipped that chapter. Briefly, you need to know what programs are right for what sort of data, and what the limitations are. So an understanding of data analysis will enable you to plan more effective experiments. Remember that the computer will be quite content to process your figures perfectly inappropriately, if that is what you request! It may also be helpful to know how to interpret the final printout – correctly.

Back to the subject of notation. As I have just pointed out, we are going to be involved in quite unsophisticated "number-crunching" and the whole point of notation is to remind us of the **order** in which we do this. Notation may look formidable, but it really isn't. It would be quite dangerous, when you start out, to think that notation enables you to do the calculations without previous experience. You just can't turn to page 257, for example, of a statistics book and expect to tackle something like:

$$\frac{(\sum xy)^2}{\sum x^2 + \sum y^2}$$

without the necessary homework on the notation! Incidentally, Box 2.1 translates this algebraic notation into English for two columns of paired numbers (values of x and y). As you progress in statistics, each part of the formula above will ring a bell in your brain for a less algebraic form of shorthand:

$$\frac{(\text{sum of cross products})^2}{\text{sum of squares for } x + \text{sum of squares for } y}$$

These terms will probably mean nothing to you at this stage, but being able to calculate the *sum of squares* of a set of numbers is just about as common a procedure as calculating the mean.

We frequently have to cope with notation elsewhere in life. Recognize 03.11.92? It's a date, perhaps a "date of birth." The Americans use a different notation; they would write the same birthday as 11.03.92. And do you recognize C_m? You probably do if you are a musician – it's notation for playing together the three notes C, E_b, and G – the chord of C minor (hence C_m).

In this book, the early chapters will include notation to help you remember what statistics such as *sums of squares* are and how they are calculated. However, as soon as possible, we will be using keywords such as *sums of squares* to replace blocks of algebraic notation. This should make the pages less frightening and make the text flow better – after all, you can always go back to an earlier chapter if you need reminding of the notation. I guess it's a bit like cooking? The first half dozen times you want to make pancakes you need the cookbook to provide the information that 300 ml milk goes with 125 g plain flour, one egg, and some salt, but the day comes when you merely think the word "batter"!

Notation for calculating the mean

No one is hopefully going to baulk at the challenge of calculating the mean height of five people – say 149, 176, 152, 180, and 146 cm – by totalling the numbers and dividing by 5.

In statistical notation, the instruction to total is "$\sum$," and the whole series of numbers to total is called by a letter, often "x" or "y."

So "$\sum x$" means "add together all the numbers in the series called "x," the five heights of people in our example. We use "n" for the "number of numbers," 5 in the example, making the full notation for a mean:

$$\frac{\sum x}{n}$$

However, we use the mean so often, that it has another even shorter notation – the identifying letter for the series (e.g. "x") with a line over the top, i.e. $\bar{x}$.

3
Summarizing variation

Chapter features

Introduction

Life would be great if the mean were an adequate summary statistic of a series of numbers. Unfortunately it is not that useful! Imagine you are frequently buying matchboxes – you may have noticed they are labeled something like "Average contents 48 matches" (I always wonder, why not "50"?). You may have bought six boxes of "Matchless Matches" to verify their claim, and found it unassailable at contents of 48, 49, 49, 47, 48, and 47. When you switch to "Mighty Matches" you equally cannot fault the claim "Average contents 48 matches," with six boxes containing respectively 12, 62, 3, 50, 93, and 68. Would you risk buying a box of "Mighty Matches"? The mean gives no idea at all of how frequently numbers close to it are going to be encountered. We need to know about the variation to make any sense of the mean value.

The example of human heights I used in Chapter 2 straightaway introduces the inevitability of *variation* as soon as we become involved in biological measurements. Just as people vary in height, so lettuces in the same field will vary in weight, there will be different numbers of blackfly on adjacent broad bean plants, our dahlias will not all flower on the same day, a "handful" of lawn fertilizer will be only a very rough standard measure, and eggs from the same farm will not all be the same size. So how to deal with variation is a vital "tool of the trade" for any biologist.

Now there are several ways we might summarize variation of a set of numbers, and we'll go through them using and explaining the relevant notation. Alongside this in text boxes (which you can skip if you don't find them helpful) we'll do the calculations on the two samples above from the different matchbox sources (differentiating them as x for "Matchless" and y for "Mighty," so both $\bar{x}$ and $\bar{y}$ are 48, but with very different variation).

Different summaries of variation

Range

"Matchless" series x had the range 47–49, contrasting with the much wider range of 3–93 for "Mighty" y. Although the range clearly does distinguish these two series of matchboxes with the same mean number of matches, we have only used two of the six available numbers in each case. Was the 3 a fluke? Or would it really turn up about one in six times? We really could do with a measure of variation which uses **all** the numbers we have collected.

Total deviation

To make the best use of the numbers in a sample, we really need a measure of variation which includes all the numbers we have (as does the mean). However, just adding the six numbers in each series will give the identical answer of 288 (6 × mean of 48).

The clue to getting somewhere is to realize that if all numbers in one of our matchbox series were identical, they would all be the mean, 48. So if the numbers are not the same, but vary, each number's contribution to total variation will be its *deviation* (difference) from the mean. So we could add all the differences of the numbers from the mean (see Box 3.1 for the calculations that apply for our matchbox example). In notation this is $\sum(x - \bar{x})$. Ignoring whether the difference is + or −, the total deviation is only 4 for "Matchless" compared with 162 for "Mighty."

This looks good. However, there is a major snag! Total deviation will just grow and grow as we include more numbers, so only samples of the same size can be compared. It would be better if our measure of variation were independent of sample size, in the same way as the mean is – we can compare means (e.g. heights of men and women) even if we have measured different numbers of each.

Mean deviation

The amount of variation per number included:

$$\frac{\sum(x - \bar{x})}{n}$$

BOX 3.1

For "Matchless," $\sum(x - \bar{x}) = (48 - 48) + (49 - 48) + (49 - 48) + (47 - 48) + (48 - 48) + (47 - 48) = 0 + 1 + 1 + (-1) + 0 + (-1) = 0$ or 4 if we ignore signs.

For "Mighty," $\sum(y - \bar{y}) = (12 - 48) + (62 - 48) + (3 - 48) + (50 - 48) + (93 - 48) + (68 - 48) = -36 + 14 + (-45) + 2 + 45 + 20 =$ again 0, but 162 if we ignore signs.

If we ignore signs, then series x has $\sum(x - \bar{x})$ (called the sum of *deviations* = differences from the mean) of 4, hugely lower than the $\sum(y - \bar{y})$ of 162 for the obviously more variable series y.

BOX 3.2

Mean deviation for the matchbox example:
For "Matchless" (x series) mean deviation is total variation/6 $= 4/6 = 0.67$
For the more variable "Mighty" (y series) mean deviation is $162/6 = 27$

---

is the obvious way around the problem. The **mean** (average) deviation will stay much the same regardless of the number of samples. The calculations in Box 3.2 give us the small mean deviation of 0.67 for "Matchless" and the much larger 27 for "Mighty."

There's not a lot to criticize about *mean deviation* as a measure of variability. However, for reasons which come later in this chapter, the standard measure of variation used in statistics is **not** *mean deviation*. Nonetheless, the concept of "mean deviation" brings us very close to what **is** that standard measure of variation, the *variance*.

Variance

Variance is very clearly related to mean deviation which I'll remind you was:

In words: $\dfrac{\text{the sum of all the deviations from the mean}}{\text{the number of numbers}}$

or in notation $\dfrac{\sum(x - \bar{x})}{n}$

> **BOX 3.3**
>
> For "Matchless" series x, calculating variance thus involves squaring the 6 deviations (see Box 3.1) from the mean, adding these squares together, and dividing by 5 instead of 6:
>
> $$\frac{0^2 + 1^2 + 1^2 + 1^2 + 0^2 + 1^2}{5} = 0.8$$
>
> and for "Mighty" series y:
>
> $$\frac{36^2 + 14^2 + 45^2 + 2^2 + 45^2 + 20^2}{5} = 1155.7$$
>
> The variances for the two series are hugely different, that for series y being nearly 1500 times the greater!

Variance is nearly the same, but with two important changes arrowed and in bold capitals:

$$\frac{\text{the sum of all the } \textbf{SQUARED} \text{ deviations from the mean}}{\textbf{ONE LESS THAN} \text{ the number of numbers}} \quad \text{or in notation} \quad \frac{\sum (x - \bar{x})^2}{n-1}$$

Variance is therefore the mean (using $n-1$) SQUARED deviation, and the calculations for our matchbox example are given in Box 3.3. The variance of only 0.8 for "Matchless" contrasts with the much larger 1155.7 for the more variable "Mighty" boxes.

Two pieces of jargon which will from now on crop up over and over again are the terms given to the top and bottom of the variance formula. The bottom part $(n - 1)$ is known as the *degrees of freedom*. The top part, $\sum (x - \bar{x})^2$, which involves adding together the squared deviations, should be called the "sum of squares of deviations," but unfortunately this is contracted to *sum of squares*. Why is this unfortunate? Well, I'll come to that at the end of this chapter but, believe me now, it is **essential** for you to remember that *sum of squares* in statistics is the technical term for **summed .. squared DEVIATIONS from the mean**. So if we are going to add any other squared numbers, we'll be careful in this book to stick with the words "adding," "addition," etc. to distinguish it from ***sum*** *of squares*.

Variance is such a frequent piece of number-crunching that it pays to lodge firmly in the recall center of your brain that variance is:

$$\frac{\text{the sum of squares (of deviations from the mean)}}{\text{degrees of freedom}}$$

Why $n - 1$?

It certainly isn't obvious that we should use $n - 1$ rather than n, and if you like you can just accept it as part of a magic formula and not worry about it! If that is your inclination, it could save you a lot of heartache to just skip to the next heading ("Why the squared deviations?") now!

However, *degrees of freedom* crop up often in statistical calculations, and are not always just one less than the total number of numbers. So it is perhaps worth making an effort to understand the concept of *degrees of freedom*. The two important basics of the concept are (i) that we are calculating statistics from a sample rather than the entire population of numbers and (ii) that the total of the numbers in the sample is used to calculate the mean which is then used as the basis for the deviations that contribute to the sum of squares.

In ignorance of the true mean of the population, we are forced into using the total of our sample to calculate a mean from which the deviations are then calculated. It is that use of the sample mean which restricts our freedom from n to $n - 1$. If we used the true mean of the population and not the mean of the sample in calculating sum of squares (of deviations from the mean), we would divide the sum of squares by n of the sample – not $n - 1$. I can remember a particular brand of pocket calculator in the 1980s which used n rather than $n - 1$ in its built-in variance program – it made a surprising difference; everyone got the wrong answers!

It is hard to find an analogy for "degrees of freedom," but looking at our bookcase at home has given me an idea. Goodness knows how many books there are on the seven long shelves, or what the mean thickness of all the books is. So, merely to try and explain degrees of freedom, lets take a rather ludicrously small sample of just six books. Well, their combined thickness is 158 mm, giving us a mean book thickness in our sample of 26.3 mm. Now, to calculate the "sum of squares" of deviations from this mean of 26.3 mm, I need to know the individual deviation from this for each of the six books. So I pick up one book (1 degree of freedom) – its thickness is 22 mm. The remaining five books must total 136 mm in thickness. I pick up another of the books (2nd degree of freedom). It is 24 mm thick.

By the time I have picked up the fifth book (5th degree of freedom) their combined thickness has reached 129 mm. There are no more degrees of freedom for my six books! I don't need to pick up and measure the last book! The total thickness of 158 mm for the six books (which is where the mean of 26.3 mm used to calculate the sum of squares comes from) tells me the last book has to be 29 mm thick. So, given the mean of a sample, I know the total – and from this the individual thicknesses of all six books after measuring only five (hence 5 degrees of freedom for a sample of six!).

By using the sample mean as the base from which to calculate the deviations for the sum of squares, we have lost one "degree of freedom." Thus variance is not the simple mean of the squared deviations, it is the squared deviation per **opportunity for variation** once we have been given the sample mean.

Why the squared deviations?

The first person who tried to teach me statistics rationalized the squaring as a way of making minus deviations positive! I guess he must have been winding us up? Why not just ignore the sign and stick with *mean deviation*? After all, once the deviations have been squared, the units no longer make sense! In Box 3.3 the variance of the "Mighty" matches was calculated as 1155.7. But 1155.7 what? Squared matches?

However, if we square root the variance, we get back to recognizably normal and unsquared matches, in this case 34.0 of them. This square root of the variance is called the *standard deviation*. You need to know that this is a uniquely important statistic; below I refer to it as "the key to the statistical lock." Look at the curve (see Fig. 5.1) of how numbers in samples from a single population are often distributed, symmetrically on either side of the mean. The curve has three identifiable points. Firstly, the highest point (the mean). Then, equally spaced on either side are the arrowed points where the line, descending as a convex curve from the mean, changes to a con-cave (hollow) curve. These identifiable arrowed points show a plus/minus distance from the mean measured by the *standard deviation*. However, we cannot calculate it directly from differences from the mean of unsquared numbers, e.g. of matches, for it depends on large differences from the mean being given more importance than small differences. So we square and sum the deviations from the mean and divide by $n - 1$ to get variance, and **only then** square root the result to arrive at what we are really after, which is the standard deviation. Like the mean, we use variance and standard deviation so often that their notation has been simplified from a formula to just "s" for standard deviation and (appropriately) "s^2" for variance.

The *mean* was our first, the *variance* our second, and now the *standard deviation* is our third important statistic that we calculate from a sample of numbers. We summarize the whole population of numbers from which our sample was drawn by . . . the *mean ± standard deviation*. For the contents of "Mighty" matchboxes, this is 48.0 ± 34.0 matches Such a summary contains information far more complete than that provided by any other calculation such as mean deviation. Assuming our sample is adequately representative of a whole population of numbers which is "normally distributed" (see later, page 26 and Fig. 5.1), the mean ± standard deviation is as good as having all the numbers, including those we didn't sample (which is usually most of them!). To put it another way, if you were to give me the mean and standard deviation of a sample of 50 eggs from an egg farm producing 500,000 eggs a week, I could work out for you the likely weight of each of the 499,950 eggs you didn't weigh! Thus I could tell you how many of those eggs would weigh exactly 65 grams, or less than 55 grams, or between 52 and 56 grams or any other statistic you would like. Hard to believe? If so, don't go away; read on!

The standard deviation

Standard deviation is the square root of variance. But what is so important about it is it is the "key" to the statistical "lock." You will find that very many statistical tests rely on calculating the standard deviation of a sample.

We calculated the mean and standard deviation of the number of "Mighty" matches per box as 48.0 ± 34.0. Well, 48.0 ± 34.0 suggests variation between 14 and 82 matches per box, but if you go back to the start of this chapter you'll see those limits only include three of the six matchboxes sampled. Well, six boxes is far too small a sample for reliable statistics given such high variability. Yet I can tell you the range 14–82 should really have included four, which is pretty close for such a tiny sample, seeing that you can't have half a matchbox and that the box with 12 was only just excluded. Are you getting the impression that I know what proportion of the matchboxes should lie within the limits of mean ± 1 standard deviation? Perhaps surprisingly, I do! Then you might guess that I also know the proportion that should lie in the ranges: mean ± 2 standard deviations, mean ± $\frac{1}{2}$ standard deviations, or even mean ± 0.35 standard deviations!

So to visualize the power of the standard deviation we can imagine a ruler with equal divisions, not marked out with any particular numbers, but with a scale of ± "**standard deviations worths**" with the mean at the center (Fig. 3.1). Any actual numerical measurement (such as numbers

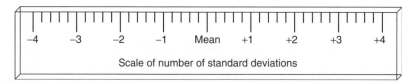

Fig. 3.1 An unusual ruler! The familiar scale increasing left to right has been replaced by units of plus or minus "standard deviation's worths" with the mean at the center.

BOX 3.4

Assume a mean of 50 with a standard deviation (s) of 10. On a scale of standard deviations worths, selected numbers would translate as follows:

23:	$23 - 50 = -27$	$-27/10 = $ **−2.7** s
31:	$31 - 50 = -19$	$-19/10 = $ **−1.9** s
50:	$50 - 50 = 0$	**mean**
55:	$55 - 50 = +5$	$+5/10 = $ **+0.5** s
74:	$74 - 50 = +24$	$+24/10 = $ **+2.4** s

---

per matchbox) can be converted to this scale (Box 3.4). As hinted above, the beauty of the **standard deviations worths** scale is that we can make predictions about the proportion of individual numbers that will fall between any two marks on the scale.

So any measurement can be expressed either as a number, **or** as a deviation from the mean **or** as a standard error worth's difference from the mean. For the moment, please just accept blindly that this is so – hopefully you will be more convinced once you have read Chapter 5, where this idea – which is so fundamental to understanding statistics – is explored in more detail.

Before moving on, it is worth just mentioning that standard deviation was once called *root mean square deviation*. This was really a much more helpful name, as it was a constant reminder in reverse ("deviation, square, mean, root") of the computational procedure, where "mean" is in $n - 1$ form. You take the deviations, square them, work out the $n - 1$ mean, and then square root the result:

$$\sqrt{\frac{\sum (x - \bar{x})^2}{n - 1}}$$

The next chapter

Often, after the relevant theory has been developed, short-cut methods have been found to speed up calculations. The trouble is that the formulae for these newer computational procedures do not carry the information content of what one is actually trying to achieve statistically. In this chapter we have used the notation $\sum(x - \bar{x})^2$ for the sum of squares (of deviations from the mean) – and this notation describes what one is doing. The next chapter is a digression describing a quicker way of arriving at the same numerical answer – but beware, although it involves adding together squared numbers, these numbers are NOT the deviations from the mean. So there has to be an additional step – a "correction" factor – before the right answer is obtained. It is really well worth converting to this new and quicker method for calculating the sum of squares of deviations – but you must always remember you are using a different formula to arrive at the same answer – **summing** the squares of deviations from the mean will not be part of the new procedure.

Spare-time activities

1 What is the variance of the following numbers, which total 90?

9, 10, 13, 6, 8, 12, 13, 10, 9

2 Express the following set of samples as their mean ± standard deviation:

1, 3, 2, 6, 4, 4, 5, 7, 6, 4, 4, 5, 3, 5, 3, 2

How many standard deviations from the mean would an observation as large as 8 represent?

4

When are sums of squares NOT sums of squares?

Chapter features

Introduction

In Chapter 3 we learnt that *variance* is the *sum of squares (of deviations from the mean)* divided by the *degrees of freedom*. This "sum of squares," $\sum(x-\overline{x})^2$ in our notation, is therefore the top half of the variance calculation. It takes a long time to subtract each number from the mean, and a quicker way has

been found of arriving at the same answer. 🐘 But never forget, you **still have to divide** by $n-1$ to convert sum of squares to variance!

My first years in research **did** involve squaring the deviations from the mean. We had to do this by mental arithmetic, by looking up tables of the squares of numbers, or by using logarithm$_{10}$ tables. In the last named method, you looked up the log. of your number, doubled it, and then reconverted with antilog. tables! It was therefore a bonus that deviations tended to be reasonably small numbers!

Calculating machines offer a quicker method of calculating sums of squares

Added squares

With calculating machines came the power to square large numbers as easily as small ones, and so the calculation of sums of squares could be speeded up by eliminating the step of subtracting each number from the mean. We can now just square and add the numbers themselves (REMINDER: I am using the word "add" instead of "sum" to, probably more in hope than in anticipation, prevent you from getting confused between "adding the squares of the original numbers" and "summing squares of deviations"). In notation, adding the squares of the numbers is $\sum x^2$. As numbers which differ from the deviations are now squared and added, we clearly cannot get at the same answer without further work. $\sum x^2$ just cannot be equal to $\sum (x - \bar{x})^2$, unless the mean is zero!

The correction factor

To turn added squares of the numbers into the sum of squared deviations from the mean, we have to subtract a **correction factor** from the added squares. This factor is the square of the total of the numbers, divided by how many numbers have been squared and added – in our notation:

$$\frac{(\sum x)^2}{n},$$

making the full notation for *sum of squares* by this "calculator" method:

$$\sum x^2 - \frac{(\sum x)^2}{n}.$$

This formula is what is called an "identity" for $\sum (x - \bar{x})^2$, since it is a different calculation which gives the identical answer. There is quite a complex algebraic proof that this is so. You will find this proof in most larger statistical textbooks, but you will hopefully be satisfied with Box 4.1, which compares the sum of squares for "Mighty" matches (see start of Chapter 3) calculated by both methods.

Avoid being confused by the term "sum of squares"

Because both computations involve adding squared values, you will appreciate there is serious room for confusion. Just keep in mind the contrast between:

summing squared deviations and the calculation is **finished**
versus
adding squared numbers but the calculation is **not** finished, you have to subtract the correction factor to **finish**.

BOX 4.1

$$\sum(x - \bar{x})^2 = (12 - 48)^2 + (62 - 48)^2 + (3 - 48)^2$$

$$+ (50 - 48)^2 + (93 - 48)^2 + (68 - 48)^2$$

$$= 36^2 + 14^2 + 45^2 + 2^2 + 45^2 + 20^2 = \mathbf{5946}$$

$$\sum x^2 = 12^2 + 62^2 + 3^2 + 50^2 + 93^2 + 68^2$$

$$= 19{,}770$$

$$\left(\sum x\right)^2 = 298^2 = 82{,}944$$

$$\left(\sum x\right)^2 / n = 82{,}944/6 = 13{,}824$$

$$\sum x^2 - \left(\sum x\right)^2 / n = 19{,}770 - 13{,}824 = \mathbf{5946}$$

It can't be repeated too often, either method gives you the *sum of squares (of deviations from the mean)* calculation; you still have to divide by $n - 1$ to convert this to *variance*.

In future in this book, we will be using the calculator method of adding squared numbers and subtracting the correction factor; just never forget this is a replacement for summing squared deviations from the mean, i.e.

$\sum(x - \bar{x})^2$. ⬛ Never forget – adding squares **never, ever,** gives you the "sum of squares" unless the numbers you are squaring are *deviations from the mean.*

Summary of the calculator method of calculating down to standard deviation

1 $(\sum x)^2$: Add the numbers together and square this total.

2 $(\sum x)^2/n$: Divide this squared total by the number of numbers in the total above – this is your **correction factor**.

3 $\sum x^2$: Square the numbers and add the squares together – this is your **added squares**.

4 $\sum x^2 - (\sum x)^2/n$: Subtract your correction factor from your added squares – this is your **sum of squares (of deviations from the mean)**, $\sum(x - \bar{x})^2$. Under no circumstances can this be negative! If your correction factor is larger than your added squares, you've made a mistake somewhere.

5 $\sum (x-\bar{x})^2/(n-1)$ or s^2: Divide the sum of squares by degrees of freedom $(n-1)$ to get your **variance**.

6 $\sqrt{[\sum(x-\bar{x})^2/(n-1)]}$ or s: Square root the variance to get the **standard deviation**.

Spare-time activities

1 Recalculate the variances of the numbers given in "Spare-time activities" 1 and 2 from Chapter 3, now using the method described in Chapter 4, and check that you get the same answer!

2 Calculate the mean and standard deviation of the following set of figures:

2.47, 2.55, 2.51.2.39, 2.41, 2.47, 2.44, 2.50, 2.46, 2.55,

2.51, 2.32, 2.50, 2.54, 2.51.

Describe 2.43 and 2.99 in terms of ± standard deviations from the mean.

5

The normal distribution

Chapter features

Introduction

In Chapter 3, I made some remarkable claims for the power of the statistics *mean ± standard deviation*, but did add the rider . . . "assuming our sample is adequately representative of a whole population of numbers which is normally distributed." So now it's time to explain what is meant by a "normal distribution," and then to try and convince you that the standard deviation is indeed the "key" to the statistical "lock" as claimed in Chapter 3.

Frequency distributions

The word "distribution" is short for "frequency distribution," i.e. the frequency of occurrence of the different numbers in a population. The *normal distribution* is a particular pattern of variation of numbers around the mean. It is symmetrical (hence we express the standard deviation as ±) and the frequency of individual numbers falls off equally away from the mean in both directions (Fig. 5.1). In terms of human height, progressively larger and smaller people than the average occur symmetrically with decreasing frequency towards respectively giants or dwarfs. What is important about this distribution is not only that this kind of natural variation

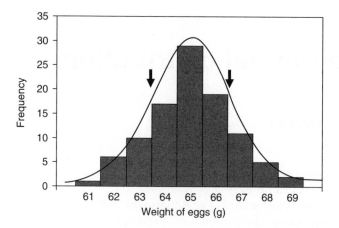

Fig. 5.1 An example of the normal distribution: the frequencies in which 100 eggs occur in different weights. The arrows show the points of inflexion where the curve changes from convex to concave on either side of the mean (giving the limits for mean ± 1 standard deviation).

BOX 5.1   Data of Fig. 5.1

It would take a lot of space to write out all 100 weights, but (starting at the lightest end) just 1 egg weighed in at 61 g, there were 6 eggs weighing 62 g, 10 at 63 g, 17 at 64 g, and 29 at 65 g, the most frequent weight. The frequencies then declined again, with 19 at 66 g, 11 at 67 g, 5 at 68 g, and the heaviest 2 eggs at 69 g.

---

often occurs, but also that it is the distribution which comes with the best statistical recipe book for data analysis and testing our hypotheses.

The normal distribution

Figure 5.1 uses the data of a sample of 100 chickens' eggs (Box 5.1), weighed to the nearest gram (g), from a large commercial poultry enterprise. The mean weight of the sample was 65.1 g.

The histogram of Fig. 5.1 shows that this clearly coincides with the most frequent weight, with eggs increasingly heavier or lighter in weight falling off in frequency roughly equally to either side of the mean. The smoothed-out curve which has been superimposed is a typical "normal distribution" curve; it is symmetrical about the mean, and is made up of two "mirror

---

BOX 5.2   Calculation of standard deviation [s] of sample of 100 eggs of total weight 6505 g

From now on, we'll use the quicker calculator method for sum of squares (of deviations from the mean), which saves us calculating those very deviations; see end of Chapter 4.

The correction factor is $(\sum x)^2/n = 6505^2/100 = 423{,}150.2$

The added squares of the numbers, $\sum x^2$, is $61^2 + 62^2 + 62^2 + 62^2 + 62^2 + 62^2 + 62^2 + 63^2 + \cdots\cdots + 69^2 + 69^2 = 423417$

So sum of squares of deviations, $\sum (x - \bar{x})^2$ = added squares − correction factor = $423{,}417 - 423{,}150.2 = 266.8$

Then variance, $\sum (x - \bar{x})^2 /n - 1$ or $s^2 = 266.8/99 = 2.7$ and the square root of this, the standard deviation (s), is 1.6.

Note: Where many numbers in the frequency distribution share the same value, as in this example, the sum of squares can more easily be calculated from the frequencies (see the addendum at the end of this chapter).

---

image" S-shaped curves. These curves give *points of inflexion* (the arrows on Fig. 5.1) where convex around the mean changes to convex towards the extremes.

The sample included no eggs lighter than 61 g or heavier than 69 g. But this was a sample of just 100 eggs from a pool of hundreds of thousands produced on the same day. So it is very likely that both lighter and heavier eggs were part of the day's production, but too rare to be included by chance in a sample of 100 eggs. The tails of the two mirror-image S-shaped curves of the theoretical normal distribution curve approach the zero frequency baseline so slowly they never actually reach it, so there is a remote chance (theoretically) of an egg weighing a kilogram – pity the poor chicken!

The mean weight and standard deviation of the sample data in Fig. 5.1 is 65.1 ± 1.6 g (Box 5.2).

I can now predict that 68% of all the eggs produced by the poultry farm, the same day as the sample was taken, would weigh between 63.5 and 66.7 g, i.e. the mean (65.1) ± 1 standard deviation (1.6 g) (Box 5.3).

What per cent is a standard deviation worth?

The prediction is possible because, for a perfect "normal" distribution, the mean and standard deviation enable us to replace the frequency histogram of the distribution with the kind of curve drawn in Fig. 5.1. We can then

BOX 5.3

What per cent of eggs in the sample of 100 fall in the range of mean $\pm$ 1s (i.e. 63.5 – 66.7 g)? All those weighing 64 and 65 g, certainly. That's 46 eggs to start with. Ten eggs weighed 63 g, so we can add 5 for the 63.5, bringing the number of eggs up to 51. Lastly we have the 66.7 for the 19 eggs weighing 66 g. 0.7 of 19 is 13, making our total eggs within the mean $\pm$ 1s limits to 64 out of a 100. Not quite the predicted 68%, but pretty close!

What per cent eggs lie outside the limits mean $\pm$ 2s? These limits are 65.1 $\pm$ 3.2 g, i.e. 61.9 – 68.3 g. The one 61 g egg and the 2.69 g are certainly outside the limits. The 0.3 of the 5.68 g eggs is 1.5, making a total of $1 + 2 + 1.5 = 4.5$, very close to the predicted 5%.

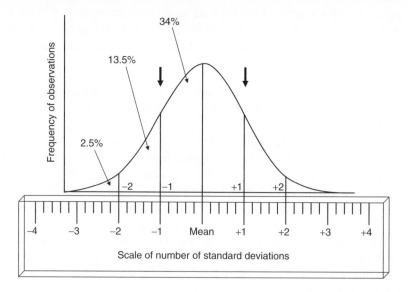

Fig. 5.2   The normal distribution curve of Fig. 5.1 with the egg weight scale replaced by the Fig. 3.1 "ruler", divided into 1 standard deviation divisions on either side of the mean. The per cent observations (eggs in this example) contained in each division of the graph is shown. This percentage division will apply to any set of observations whose frequencies follow a normal distribution.

line up our scale of "standard deviation worths" so that the mean coincides with the peak of the distribution curve and the $\pm$1s points line up with the points of inflexion of the curve (see earlier). When we do this (Fig. 5.2), the data become divided into blocks of different proportions. 34% lie between the mean and 1s in either direction – hence my 68% prediction between mean $\pm 1s$. 2.5 % of the data at either end are more extreme than $2s$ distant from the mean (Box 5.3), leaving 13.5% in each of the two bands between 1 and $2s$ distant.

Are the percentages always the same as these?

Well, yes, actually – they are, in perfectly symmetrical normal distributions. You may find this hard to believe, and you'll either have to take my word for it or get yourself a more advanced text and follow the mathematical arguments! Something that may help to convince you is to take the egg distribution curve, which seems to fit the percentage frequency pattern I am trying to convince you about, and imagine you have "modeled" it in three colors of plasticine (dark, lighter, and white, Fig. 5.3 – bottom left). However we distort the distribution, vertically or horizontally – as long as we keep it symmetrical, the three colors remain in their original proportions and the equal spacing of the standard deviation scale on the horizontal axis is retained. All we are changing is the distance between the junctions where the plasticine changes color and the peak height of the curve. Has that convinced you?

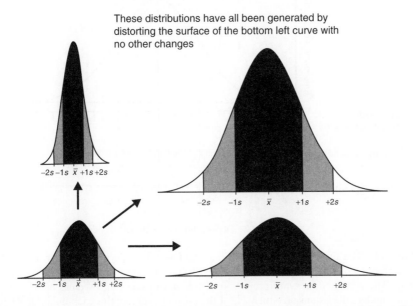

Fig. 5.3 Demonstration that the areas under the normal curve containing the different proportions of observations maintain their proportional relationships in spite of distortion of the curve, provided it remains symmetrical about the mean.

Other similar scales in everyday life

It should not be a very difficult concept to grasp, for we use such "relative" scales all the time in everyday life, whenever proportions seem more important than absolute values. When I ran a Scout troop and was chuffed that I got a 60% turnout at church parade, that meant 22 boys, yet the much larger number of 1100 people is a disappointing 30% turnout at our local elections. When I drive into work I'm "nearly there" as I turn into the University entrance in Reading; yet when I drive back to Reading from Grimsby I would say I'm "nearly home" at Maidenhead, still 14 miles away.

Figure 5.4 is another demonstration of the unifying concept that the normal distribution provides for all suitable data, and plots three scales against the per cent frequencies of the normal distribution curve. The top scale is standard deviation worths, the second is the distribution of the numbers of "Mighty" matches in boxes (from Chapter 3), and the bottom scale is the weight of eggs from this chapter. Note that, although all three

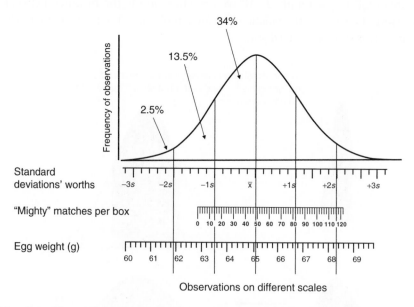

Observations on different scales

Fig. 5.4 The unifying concept of the scale of standard deviations for two sets of observations of very different ranges of 0–130 for matches per box and 61–69 g for eggs. The vertical downward projections divide both ranges into sectors of 0–1 and 1–2 standard deviations on either side of the mean; to achieve this each major division on the matches scale represents 20 matches, but only 1 g on the egg weight scale. The scales therefore have to be at different "magnifications" in order for both to fit the standard deviations scale.

scales have been made to fit the same normal curve, that for matches covers over 130 matches while for eggs the whole scale is only about 8 grams.

The standard deviation as an estimate of the frequency of a number occurring in a sample

The magnitude of any individual observation, be it 23 "Mighty"matches, a 63 g egg, a man 160 cm tall, or 10 grasshoppers per m^2 of grassland, is some unit of standard deviation worths away from the mean for those particular objects or organisms. The more standard deviation worths the observation is away from the mean, the rarer is its frequency in the population of possible numbers available for sampling. So numbers within 1s of the mean are pretty frequent; we could describe them as "quite usual." Numbers between 1 and 2s from the mean are "a bit on the large or small side." Beyond 2s distant from the mean, the numbers verge on the "exceptionally large or small." The units themselves do not matter. No one has the problem that a "normal" person is considerably larger than an "exceptionally large" dachshund! Yet the first is within 1s of the mean whereas the second is more than 2s larger than its mean.

From per cent to probability

133 "Mighty" matches in a box or a 61 g egg appear to have little in common, but Fig. 5.4 reminds us that they have something very much in common. They are both 2.5s away from the mean, and so are both equally unusual matches or eggs, though the number of matches is exceptionally large and the egg is exceptionally small. The chance of picking either in a random grab from a pile of matchboxes or eggs is the same, less than a 2.5% (= 1 in 40) chance.

In statistics, we write this as "$P < 0.025$." P stands for "probability" where $P = 1$ is 100% certainty.

The standard deviation is perhaps best regarded as a unit of "percentage expectation." We can add this "expectation of occurrence in a sample" scale to the standard deviations worths ruler (Fig. 5.5.). In biology, the convention is to regard a datum more than $\pm 2s$ away from the mean, which therefore has a less than 1 in 20 chance of "turning up" in a sample, as an unlikely event with $P < 0.05$. 2.5 of this 5% probability comes from data beyond the 47.5% point at either end of the "percentage included" scale. We know that 63 g eggs do occur on our egg farm. However, if we picked up just one egg and found it was so exceptionally small, we might

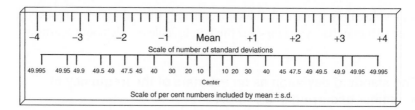

Fig. 5.5 A scale of "per cent observations included by different standard deviations worths" added to the "standard deviations worths" ruler of Fig. 3.1.

suspect it had come from a different establishment. And that, in a nutshell (or eggshell!) is the basis of decision-making in biological statistics.

Addendum on calculating sum of squares from a frequency distribution (using the egg data from the start of this chapter).

Datum egg wt (x)	Frequency (f)	f multiplied by x (fx)	fx multiplied by x (fx^2)
61	1	61	3,721
62	6	372	23,064
63	10	630	39,690
64	17	1,088	69,632
65	29	1,885	122,525
66	19	1,254	82,764
67	11	737	49,379
68	5	340	23,120
69	2	138	9,522
Totals	100	6,505	423,417

	Number eggs	Total egg weight	Added squares
What these totals represent in the formula for variance	n	$\left(\sum x\right)$	$\sum x^2$

$$\text{Sum of squares} = \sum x^2 - \left(\sum x\right)^2 / n = 423{,}417 - \frac{6505^2}{100}$$

$$= 423{,}417 - 423{,}150.2 = 266.8$$

Answer checks with that in Box 5.2

EXECUTIVE SUMMARY 1
The standard deviation

Just about all the statistical calculations we will handle in this introductory book involve summarizing the variability of a set of numbers as one single number (or statistic) – based on how different individual numbers in a set are from the mean (= average of the whole set). This difference between a number and the mean is called the **DEVIATION**.

The average deviation would be:

$$\frac{\text{Sum (number} - \text{mean)}}{\text{Number of numbers}}$$

The **VARIANCE** (the measure we actually use to summarize the variability of numbers) is a slight modification of this:

$$\frac{\text{Sum (square of \{number} - \text{mean\})}}{\text{Number of numbers} - 1}$$

An easier way on a calculator to get the top half of the expression (called the **SUM OF SQUARES** – it is actually the **sum of squared deviations**!) is to calculate **add the squares** and subtract a **correction factor**.

The **added squares** are obtained simply by adding the squares of the numbers. The **correction factor** is obtained by squaring the total of the numbers and dividing by the number of numbers.

The **STANDARD DEVIATION** (s) is the square root of variance. It is the "key" to the statistical "lock."

Importance of the standard deviation

1 We can express any NUMBER in a new way – as the **mean** *plus or minus* **so many "standard deviation's worths"**: e.g. given a population with a mean of 20 and standard deviation of 10, the figure of 15 is also "mean $-1/2$ standard deviation (= 5)." Similarly, the number 40 is "mean $+2$ standard deviations (= 20)."

2 GIVEN A POPULATION (e.g. heights of people) with the mean somewhere in the middle, and the number of people with various heights becoming progressively fewer on either side of the mean as we tend towards giants and dwarfs, the "standard deviation" allows us to judge how frequently something more extreme than a given height is likely to occur. This is just by the very nature of populations of numbers symmetrical about the mean – it just happens to be so!

Examples

Mean $= 100$, $s = 50$. The number 180 has a deviation of $+80$, which is also mean $+1.6s$.

Mean $= 2$, $s = 1$. The number 3.6 has a deviation of only $+1.6$, but is the same in s units as the number 180 in the first example, i.e. mean $+1.6s$.

If we were to sample 100 numbers from both populations, a little over 5% would be larger than 180 in the first population; similarly the same proportion (a little over 5%) would be larger than 3.6 in the second population (you can find the actual % for any s "worths" in statistical tables).

What have the following in common?:

A 120 cm high daisy, a 100 g fully grown lettuce, a four-leaved clover and a 240 cm tall man?

They are all somewhat unusually large, small or many-leaved! Instead of using words like "normal," "a bit on the large side," "unusually small," or "extraordinarily small," it is "standard deviation's worths" which allow us to be much more precise. We might be able to say "in the largest 2.5%" (if $>$ mean $+ 2s$) rather than just "very large."

6

The relevance of the normal distribution to biological data

Chapter features

To recap

The normal distribution is defined by just two statistics, the *mean* and the *standard deviation*. It has had a "recipe book" developed for it, based on its symmetry and the percentages of areas under the distribution curve related to multiples of the standard deviation ("standard deviation's worths"). We can calculate the frequency with which different numbers will be found in an infinite population from the results of just a sample thereof.

So it is the most useful distribution for statistical analysis, but we have to admit two things straight away:

1 Biological data are frequently not normally distributed, but are asymmetrical. We often find they peak well to the left with a long thin tail of high values. This is often found with random or clumped data (the gray histogram in Fig. 6.1 shows the frequencies of larvae of a stem boring fly in 300 samples of a meter of row in a wheat crop) – common phenomena in biology. With this kind of distribution the mean does not coincide with the peak of the curve (the arrow on Fig. 6.1). Calculating a symmetrical standard deviation is plain stupid. The standard deviation should contain 34% of the observations on either side of the peak frequency (which should coincide with the mean). The true and asymmetrical spread of 34% of the data on either side of the peak frequency is shown in solid black columns in Fig. 6.1, and the position of the mean

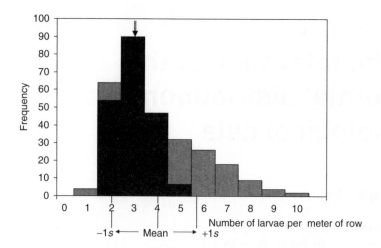

Fig. 6.1 The frequency of occurrence of larvae of the frit fly (*Oscinella frit*) in 300 samples of a meter of row in a wheat crop, showing the peak frequency not coinciding with the arithmetic mean. Normal distribution statistics calculations for mean ±1*s* give the range shown under the horizontal axis; the proportion of samples which should actually be included by peak ±1*s* (i.e. 102 samples = 34%, on either side of the peak) is shown in solid black around the peak frequency (arrowed).

and standard deviation calculated from normal distribution statistics is contrasted on the horizontal scale. The normal distribution statistics are clearly not appropriate.

2 Any estimate of the standard deviation is just that, only an estimate from a sample of the true value of the entire population. Obviously it can be way off beam if we have taken too few samples in relation to the variation in the population. Figure 6.2 shows six estimates of the standard deviation (at each of six different sample sizes) from a population with a true standard deviation of 21.4.

This chapter gives some advice on handling these two problems.

Is our observed distribution normal?

As pointed out earlier, the theoretical normal distribution has no finite upper or lower limits. Thus a population of eggs averaging 65 g could theoretically include a 1 kg egg, if there were enough eggs for an event to occur that was 621 standard deviations' worth larger than the mean. Already, by the time we have moved 3*s* from the mean, we are into the realm of pretty rare events at the level of 6 chances per thousand. By 621*s*,

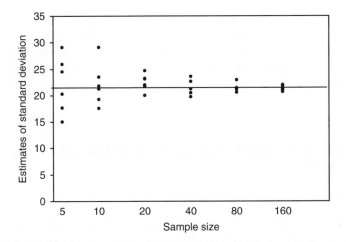

Fig. 6.2 Results of calculating the standard deviation of repeat samples of different sizes from a single population with a true *s* of 21.4 (horizontal line). Even with 10 samples, we can get estimates of this true 21.4 varying from about 17 to 30!

we are more likely to hear we've won the national lottery while hitting a hole in one!

That a chicken should lay a 1 kg egg is, in the real world, even more unlikely than the normal distribution would suggest. The world record for a chicken egg is less than 150 g, which is already 142*s* for the eggs in our calculation. There is no doubt that the tails of biological distributions are much shorter than those of the theoretical normal distribution. In practice, this makes little difference to our estimation of probabilities from the normal distribution; it merely means that events of a particular magnitude are in truth marginally less likely than our probability statistics would predict.

Checking for normality

The real problem is whether our experimental data are even an approximate fit to a normal distribution. This is easily checked with large samples. There should be roughly equal numbers of observations on either side of the mean. We may even have enough samples to draw a histogram like Fig. 6.1 (where there are 200 observations lower than the mean, but only 96 higher!) to judge the departure of the distribution from symmetry. Things are more difficult when, as usually, we have only a few samples. In experiments, it is not uncommon to have no more than three data per treatment. However, even here we can get clues. If the distribution is normal, there should be no relationship between the magnitude of the mean and its standard deviation. However if, for example, small means clearly show less variation than large means, a poor fit to a normal distribution is indicated. If the observations come from a single distribution (e.g. our egg data in Chapter 5), we can combine data at random into different-sized groups in order to obtain the range in mean values we need to do the same kind of exercise.

What can we do about a distribution that clearly is not normal?

Transformation

With large samples, we can identify whether the asymmetry causes the peak to lie below or above the arithmetic mean. A statistician may even be able to identify that the data follow a particular type of non-normal distribution. Techniques are available for changing the horizontal scale (the observed values) to a nonlinear one before statistical treatment of the data is attempted. This process of converting the observations to a function such as logarithm or square root in order to "normalize" the distribution is called *transformation*. We can use the data of Fig. 6.1 as an example. The peak is to the left of the mean and the distribution is actually close to random, which is not unusual for the distribution of the pest insect in question. A perfect random distribution is defined as having the variance of the observed values equal to the mean. Our insect distribution has a mean of 3.9 and s of 1.86. Variance (s^2) is therefore 3.46, pretty close to the mean. A random distribution can be "normalized" by converting the observed values to logarithm to base 10. Usually we first add 1 to each observed value, so that the transformation is $\log_{10}(x + 1)$. This addition of 1 enables us to have a log. for zero values. There is no logarithm$_{10}$ for zero – zero is the log. of 1!

Figure 6.3a shows the untransformed data, and Fig. 6.3b the curve after transformation. You can see how the mean of the transformed data is nearer the peak, and areas to left and right of the 1 standard deviation limits are more of similar size than these areas in Fig. 6.3a.

The subject of *transformation* is outside the scope of this book, but the *log. transformation* is particularly common. Two other common transformations are:

Square root transformation

This is used for more highly clumped data where the variance is obviously greater than the mean, but not to the extent where the reciprocal

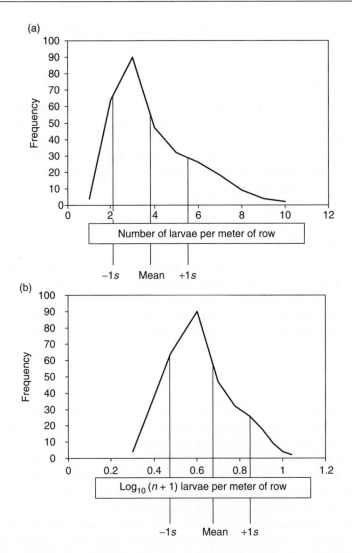

Fig. 6.3   The frequency distribution of frit fly larvae (see Fig. 6.1.) with normal distribution statistics on the horizontal scale. (a) Untransformed; (b) after logarithmic transformation.

transformation might be appropriate (see below). As there is no square root of a minus number, a constant needs to be added to each observation to make even the lowest observation positive. The transformation therefore becomes $\sqrt{x + \text{constant}}$. If there are numbers between 0 and 1, then it pays to use 1 as the constant, since the transformation of the distribution towards normality is spoilt by the phenomenon that the square root of a fraction becomes **larger** than the observation being transformed.

Arcsin or angular transformation

This is often used to normalize percentage data, especially where high or low percentages are involved. It makes little difference in the 30–70% range.

   The message here is that non-normal data can be transformed, so that they can be analyzed by the extensive "recipe book" developed for normal distributions. Once you are aware of this, the necessary further information can always be sought elsewhere.

Grouping samples

Selecting groups of four samples at random, and using the means of the groups as the data for analysis, will normalize even extremely non-normal distributions. This is because the four samples will by chance have been drawn from different parts of the distribution and the peculiarities of the distribution will have been "cancelled out" in using their mean.

Doing nothing!

With small samples, it will be hard to tell whether or not the distribution is normal. Failure to make the appropriate transformation is not the end of the world. Even in Fig. 6.3a, the fit of the mean and standard deviation to the non-normal curve is not catastrophically bad. One can also take comfort in the fact that failure to make the appropriate transformation is probably scientifically safe. Although we may fail to detect a difference that really exists between our experimental treatments, we will not be misled into claiming a difference which would be shown to be spurious by transforming the data. This is because failure to transform is more likely to overestimate the variation in the data than the appropriate transformation.

How many samples are needed?

Our calculated mean and standard deviation from a sample are only estimates of the true values for the whole population of numbers. How good these estimates are depends on how many samples we have taken and how representative they are of the population. The greater the variability of the population, the more samples will be needed to achieve the same precision.

So the answer to the question "how many samples are needed?" depends on what one is sampling! We often can't take enough samples, or replicate a treatment in an experiment sufficient times to get accurate estimates of the mean and variability. However, statisticians have produced tables (see later) which give us various forms of magnification factor for our estimates of variability from small sample sizes. Such factors prevent us drawing foolish conclusions from the application of probabilities appropriate to large samples to our more restricted sample data.

We may at times have to accept that our experiment was too small to enable us to validate an experimental result statistically. We will have wasted our time, effort, and money. However, it is an equal waste of resources to do so many repetitions of treatments that our analysis shows that many fewer would have been enough.

Factors affecting how many samples we should take

How many samples to take is therefore a question well worth asking, and three considerations will contribute to the answer:

1 The variability of our biological material – we can get an idea of this from some preliminary sampling.
2 The size of difference that we wish to detect. If, for example, we are testing a new slug killer, is a 5% reduction in leaf damage going to persuade people to buy our product? Or do we need at least a 30% reduction? A realistic aim can reduce the number of samples we need to take.
3 How important is it that, if the size difference we are looking for exists, that we actually detect it? A 100% certainty of detection usually means an impossible workload, but it pays to know what chance we have of finding our difference with the workload that is practical.

Calculating how many samples are needed

There is a way of using the above considerations to estimate the right number of samples to take, but it is not worth describing it until we have covered a lot more statistical ground. The method is therefore at the end of this book, in Appendix 1. It really is worth doing – there is no point in embarking on an experiment if it clearly cannot yield results. Also working the method will identify for us if we can control the variation, and therefore our number of samples, by a *transformation* of the data (see earlier in this chapter).

7

Further calculations from the normal distribution

Chapter features

Introduction

Things are now getting serious! This chapter is probably the most important in the book, so don't rush it. If, by the end, you really understand what the *standard error of the difference between two means* is and its importance in evaluating biological research, then statistics will hold no more fears for you – once this important central chunk makes sense, the whole subject will cease to be a "black box." And that's a promise!

So far we have explored how we can use limited sampling (from a large population) to define that whole population as a symmetrical "normal" distribution, with a mean and standard deviation as estimated from our sample.

In this chapter, we come to the reason for doing this! In a nutshell, calculations based on the *standard deviation* allow us to make objective judgments – on standards accepted by other scientists – about the validity of what we observe in biological experiments.

Is "A" bigger than "B"?

Suppose we grow large plots of two varieties of lettuce, "Gigantic" and "Enormous," which are both mature and ready for harvest at the same time. The question we want to answer is "Does one variety produce heavier lettuces than the other, or is there really no noticeable difference?" If we compare just one "Gigantic" with one "Enormous" lettuce, one is bound to be heavier than the other, yet we cannot claim that this has answered our question. We might be comparing a particularly minute specimen of "Gigantic" with an exceptionally large one of "Enormous." Common sense tells us we need several specimens of each variety before we can make a judgment.

Normal distribution statistics enable us to replace weighing a large number of lettuces of each variety with quite a small sample, and to discover if we have weighed enough lettuces to convince others as to whether the two lettuces differ in weight (see also Appendix 1).

We could take two samples (each of say four lettuces), either from the same variety or from each of two different varieties (i.e. from the same seed packet or from two different ones). In neither case is it likely that the means of our two samples of four lettuces will be identical. Looking at the difference in mean weight of the samples from the two varieties, we will want to judge whether or not this difference is merely of a size that could have equally arisen by chance when sampling the same variety twice. If the large size of the difference makes such a possibility most unlikely, we say that the difference between the means is "statistically significant."

The yardstick for deciding

As samples from a single source will always show some variation ("background variation"), we need a "yardstick" for judging the likelihood of a **difference between two means** occurring by chance sampling from such "background" variation of our biological material (i.e. in the absence of any other source of variation such as different packets of lettuce seeds!). In Chapters 5 and 6 the *standard deviation* was described as just such a "likelihood yardstick" in respect of the likelihood (= frequency) of a single observation of any given magnitude being picked from the distribution. Now we need a comparable yardstick to judge the size of a **difference between two means**. There is indeed such a yardstick: it is the *standard error of a **difference between two means*** (or s.e.d.m. for short). A complication here – "error" replaces "deviation" when talking about variation of means rather than that of individual data; sorry about that!

What the *standard error of difference between means* denotes biologically is one of the hardest concepts to grasp. Army quartermasters used to list their stock "backwards" – for example "jackets, green, soldiers, for the use of." I always think that using this approach in statistics can make things clearer: "means, differences between, standard error of." It is a really important concept and everything in this chapter so far has been nought but a preliminary to the calculation of this vital statistic.

The standard error of a difference between two means of three eggs

Using the raw (pardon the pun!) egg data from Box 5.1 together with random number (1–100) tables, I sampled 300 eggs and recorded the mean of each set of three consecutive eggs, to give me 100 means. I took these in pairs to get 50 differences between means, subtracting the weight of the second in each pair from that of the first. These 50 means are shown in Box 7.1. We can treat these like any 50 numbers and calculate their variance as 1.7, and square root this to get the "standard deviation" (but now called the "standard error") of 1.3.

BOX 7.1

50 differences between two means of the weights of 3 eggs randomly drawn from the egg weight data in Box 5.1: +0.67, 0, −0.67, −0.67, +2.00, +0.33, 0, +3.33, +1.33, +1.67, −0.67, −1.67, +1.67, −1.67, +1.33, +2.00, +1.00, −0.67, −1.33, +1.33, −1.33, +1.00, +0.67, −0.67, +1.33, +0.33, +1.00, +2.67, 0, −0.33, +1.67, +1.33, −2.67, +1.67, +1.00, −2.00, +0.33, +0.67, +0.33, +0.33, −2.33, 0, 0, +1.00, 0, −1.00, +1.00, −0.67, +0.67 and +2.33 g.

These 50 differences were a data set whose variance I could calculate as if they were single observations. In the same way as for individual eggs (page 48), we would expect the average difference to be zero, but the chances of sampling give us a small mean difference of +0.35 (from a total of +17.64).

[If you still need to look up how to calculate a variance (s^2), go back to page 15 and remind yourself that it is sum of squares of deviations divided by degrees of freedom, and then to page 23 to remind yourself of the "added squares of numbers – correction factor" technique for calculating sum of squares of deviations.]

The correction factor $((\sum x)^2/n)$ is $+17.64^2/50 = 6.22$.

The "added squares" (a minus squared is positive) are $+0.67^2 + 0^2 + (−0.67^2) + \cdots + 2.33^2 = 89.32$.

Therefore sum of squares $= 89.32 − 6.22 = 83.10$.

Variance is $83.10/(n − 1) = 83.10/49 = 1.7$.

Standard error ($=$ standard deviation when means are involved) is therefore $\sqrt{1.7} = 1.3$.

So 1.3 is our "likelihood yardstick" for judging whether a difference between two means (of egg weights each based on three eggs) is a likely or unlikely difference to sample within a single batch of eggs. But how can we work out this yardstick when we have far fewer data?

Derivation of the standard error of a difference between two means

Fortunately we can easily calculate the s.e.d.m. directly from the variability between individual samples rather than having to create lots of means to find differences between.

All we need is the variance shown by the samples and the number of samples in each of the means being compared, because the variance of means has a very simple relationship with the variance of the set of individual samples, and the variance of differences between means has a similarly simple relationship with the variance of means.

So in this chapter we will derive the *standard error of a difference between two means* from the *variance of a set of individual samples* in three steps. It will be painfully slow for anyone already familiar with the s.e.d.m., but hopefully the lengthy explanation will make it easier for beginners. If you are familiar with calculating the s.e.d.m., and just need a concise reminder, then skip the sections indicated by a vertical line in the margin which use the egg weight data to convince readers that the statements made check out with actual numbers. As I'm going to need to use the same words like variance and standard error so often, I hope it's OK with you if I make frequent use of the following abbreviations:

s for the standard deviation relevant to the variation among single data
s^2 for variance calculated from a set of single data
s.e. for standard error (relevant to means)
s.e.d.m. for the standard error of differences between means.

Step 1 – from variance of single data to variance of means

End result: The variance of means of "n" numbers is the variance of single data divided by "n" – it can't be simpler than that, can it?

Check it out: The mean ($\bar{x}$) for the sample of 100 eggs was 65.1, s^2 was 2.7, and s ($=\sqrt{s^2}$) was 1.6. To obtain several means of smaller samples, I've used random number tables to select 20 groups of 3 eggs to give me 20 mean values (Box 7.2). These 20 means can now have their s and s^2 calculated by treating them as 20 single numbers.

BOX 7.2

The first three eggs I picked at random weighed in at 65, 62, and 64 g, giving $\bar{x}$ of 63.67).

The next three were by chance heavier on average – 68, 66, and 66 again ($\bar{x}$ of 66.67, while the next group were lighter at 61, 63, and 65 ($\bar{x}$ of 63.00).

Eventually I had 20 selections with a range of means – 63.67, 66.67, 63.00, 64.67, 65.33, 66.63, 64.67, 65.00, 65.00, 65.33, 64.33, 64.00, 64.67, 64.00, 64.33, 65.67, 64.67, 66.33, 65.67, and 65.00 (the total of these 20 numbers is 1298.34 and their mean is 1298.34/20 = 64.92).

[Now go back to page 15 and remind yourself that s^2 is sum of squares of deviations divided by degrees of freedom, and then to page 23 to remind yourself of the "added squares of numbers – correction factor" technique for calculating sum of squares of deviations.]

The correction factor $((\sum x)^2/n)$ is $1298.34^2/20 = 84{,}284.34$.

The "added squares" are $63.67^2 + 66.67^2 + 63.00^2 + \cdots 65^2 = 84{,}300.74$.

Therefore sum of squares = $84{,}300.74 - 84{,}284.34 = 16.40$.

Variance is $16.40/(n-1) = 16.40/19 = 0.86$.

Compared with single eggs ($\bar{x} = 65.1$, $s^2 = 2.7$), the 20 means each of three eggs had $\bar{x} = 64.9$ and variance $= 0.86$. The similarity of the two means ($\bar{x}$) of 65.1 and 64.9 is no surprise – after all, they are both estimates of the same thing, the mean egg weight of the whole production of the egg farm. The variances are however rather different; again no surprise since one would surely expect that averaging the weights of three eggs would cancel out the variation between individual eggs to some extent, with a reduced variation as the result. But note that the variance of individual eggs is $2.7/0.86 = 3.1$ times greater than the variation of means of 3 eggs – 3.1 is remarkably close to 3? Got it? So the variance of means of three eggs (0.86) can be obtained from the variance of single eggs (2.7) by dividing by three. The variance of means of 10 eggs would be 2.7 (the variance of individual eggs)/10 $= 0.27$, that of means of 50 eggs would be $2.7/50 = 0.054$, etc. To be able to do this from the individual data contributing to the only mean we may have is a very important page in the "normal distribution statistics recipe book."

There is no problem in square rooting this variance of the 20 numbers (each being the mean of 3 eggs) to get the standard deviation as $\pm\sqrt{0.86} = \pm 0.93$ g (no, that's not an error – square roots of values less than 1 do get larger!). However (see above), we don't call it the standard deviation, but the standard error. This saves writing "of individual data" or "of the mean" behind either term.

So standard error (s.e.) is the square root of the variance of the mean, and in notation is

$$\sqrt{\frac{s^2}{n}}.$$

We can also derive the s.e. from the standard deviation (s) of the individual numbers by re-arranging the algebra of

$$\sqrt{\frac{s^2}{n}} \quad \text{as} \quad \frac{s}{\sqrt{n}}.$$

Figure 7.1 shows the scales for both individual egg weights and the mean of 3 eggs under the same normal distribution curve and thus the same scale of standard deviations'/errors' worths (the scale for the mean weight

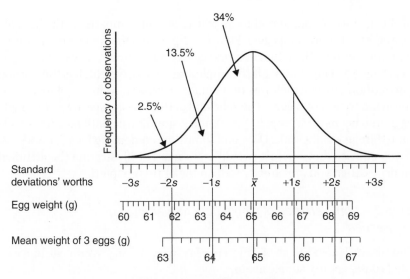

Observations on different scales

Fig. 7.1   The unifying concept of the scale of standard deviations/errors for individual observations and means for data on egg weights. The vertical downward projections divide both ranges into sectors of 0–1 and 1–2 standard deviations on either side of the mean; to achieve this each major division (1 g) on the mean egg weight scale is therefore wider than for the single egg weight scale. The scales therefore have to be at different "magnifications" in order for both to fit the standard deviations/errors scale.

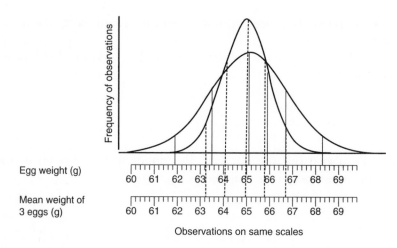

Egg weight (g)

60 61 62 63 64 65 66 67 68 69

Mean weight of
3 eggs (g)

60 61 62 63 64 65 66 67 68 69

Observations on same scales

Fig. 7.2   Figure 7.1 redrawn so that the same scale is used for weight, whether of individual eggs or means of three eggs. Now the normal curve with the ± standard deviation divisions projected down to the single egg scale with solid lines has to be redrawn as a narrower curve for means of eggs, with the ± standard deviation divisions projected down with dotted lines.

of 3 eggs has of course had to be widened to fit compared with the scale of weights of individual eggs). Figure 7.2 presents the same data, but now with the standard deviations/errors scale replaced by actual weight of eggs in grams. So in Fig. 7.2, with the single scale of egg weight, it is the normal distribution curve which has to be made narrower, to fit the statistics for the mean weight of 3 eggs. The solid and dotted lines, respectively for single eggs and means of 3, project the ±1 and 2 standard deviations/errors scale onto the egg weight scale. Do spend some time comparing Figs 7.1 and 7.2. If you can see they are just two different ways of illustrating the identical statistics, then you are already well on top of this chapter!

Step 2 – from variance of single data to "variance of differences"

End result: The variance of differences is twice as large as the variance of single data – pretty simple again.

Check it out: Apart from means, there is another set of numbers we can generate from the weights of individual eggs, and that is differences in weight between two eggs picked at random (Box 7.3). These numbers form another distribution, this time around a very small mean, since + and − differences cancel out in the adding up the total. If you think about it, the true average difference of numbers from the mean cannot be anything other than zero – that's what the mean is, it's the **middle** number!

BOX 7.3

Again using random numbers and the weights of eggs from Box 5.1, the first two eggs selected weighed 65 and 68 g. The difference in weight (second weight subtracted from first weight) was therefore −3 g.

I repeated this over and over again to get 50 such differences. Sometimes the pair of eggs were the same weight, making the difference zero. The 50 differences came out as 1 at −5 g, 3 at −4 g, 6 at −3 g, 5 at −2 g, 5 at −1 g, 11 at 0, 7 at +1 g, 2 at +2 g, 7 at +3 g, and 3 at +4 g. *You can tell by the chance few at +2 g that I really did this and haven't made the numbers up!.*

The total of these .50 numbers is zero and their mean is also zero. This exact zero is pure chance; usually there will be a small total and mean.

[If you still need to look up how to calculate a variance (s^2), go back to page 15 and remind yourself that it is sum of squares of deviations divided by degrees of freedom, and then to page 23 to remind yourself of the "added squares of numbers – correction factor" technique for calculating sum of squares of deviations.]

The correction factor $((\sum x)^2/n)$ is $0/20 = 0$.

The "added squares" (a minus squared is positive) are $-5^2 + (-4^2) + (-4^2) + \cdots + 4^2 = 278$.

Therefore sum of squares $= 278 - 0 = 278$.

Variance is $278/(n-1) = 278/49 = 5.7$.

Standard deviation is therefore $\sqrt{5.7} = 2.4$.

Box 7.3 shows that the variance of the 50 differences comes out at 5.7. Clearly 5.7 and 2.7 are close to an extremely simple relationship – that the variance of differences is twice that of individual numbers. In notation, the variance is therefore $2s^2$ (where s^2 is the variance of individual values).

The standard deviation (not standard error, as we are dealing with individual numbers) of differences between two observations is then the square root of variance (2.4 g for differences between two eggs, see Box 7.3). In notation this is $\sqrt{2s^2}$ or (the same algebraically) $s\sqrt{2}$ (if we want to derive it from s rather than s^2).

Step 3 – the combination of Steps 1 and 2; the standard error of difference between means (s.e.d.m.)

End result: From the distribution of individual values we have derived two other distributions: first that of means of n numbers, and then that of

BOX 7.4   (See also Fig. 7.3)

It's pretty obvious, really. The original egg data had 61 g as the lightest egg, a mean almost spot on 65 g and the heaviest eggs were 69 g. As deviations from the mean, we had 66 g eggs (a deviation of +1), 64 g eggs (a deviation of −1), then deviations of +2, −2, +3, −3, etc. with the largest deviations +4 (69 g) and −4 (61 g). **This is a range of 8 g.**

By contrast, with differences between pairs of eggs, the mean is 0, and we have differences of +1 (67 − 66 g eggs), −1 (65 − 64 g), +2, −2 and so until +8 (69 − 61 g) and −8 (61 − 69 g). **This is a range of 16 g.** Thus the variation of deviations from a mean of 0 of differences between eggs is twice the magnitude of the variation of deviations of individual eggs from the mean of 65 g (Fig. 7.3).

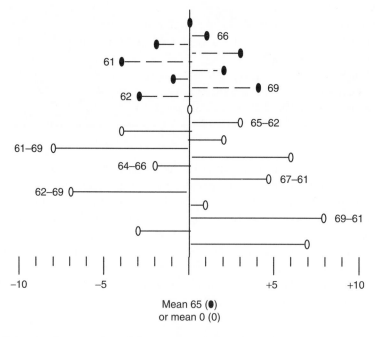

Mean 65 (●)
or mean 0 (0)

Fig. 7.3   Random sampling of the same population of numbers gives a much smaller range of deviations from the mean (65) for single observations (black ovals) than for differences (mean zero, white ovals.). See Box 7.4.

differences between two numbers. Starting with the s^2 of the individual values, it turned out that the variance for means of n numbers was s^2/n and for differences between 2 numbers it was $2s^2$. Now we can put the two together as:

$$\frac{2s^2}{n} \text{ for the } \textit{variance of difference between means,}$$

making $\sqrt{2s^2/n}$ the *standard error of differences between two means* (*s.e.d.m.*).

Check it out: The variance of differences between means is simply the combination of multiplying the variance of individual numbers by 2 and dividing by "n." The s^2 for single eggs was 2.7. Double it (for **differences**) to 5.4. The variance of **differences between means** of 3 eggs should then be 5.4 divided by 3 (for means of 3) $=1.8$. To convince you, go back to the calculation (near the beginning of this chapter, page 44) of the variance of lots of sampled differences between such means of 3 eggs (Box 7.1). This variance was 1.7, satisfyingly close (given the element of chance involved when sampling) to the variance of 1.8 we have just calculated from the variance shown by individual egg data.

As the final step – square rooting either the predicted 1.8 or the calculated 1.7 gives ±1.3 g as the **standard error of differences between means** of 3 of our eggs, which in notation is:

$$\pm\sqrt{\frac{2s^2}{n}} \quad \text{or} \quad \pm s\sqrt{\frac{2}{n}}$$

if we begin with standard deviation rather than variance.

Recap of the calculation of s.e.d.m. from the variance calculated from the individual values

Double the variance (for **differences**) of the individual values (in the egg example we double the variance of 2.7 for single eggs to 5.4). Then divide (for the **variance of differences between means**) by the number of observations in each of the means being compared (thus for means of 3 eggs we divide by 3 $= 1.8$).

As the final step – square rooting this variance gives us the **standard error of differences between means**, which in notation is:

$$\pm\sqrt{\frac{2s^2}{n}} \quad \text{or} \quad \pm s\sqrt{\frac{2}{n}}$$

(in the egg example this is $\sqrt{1.8} = 1.3$).

Coming back to the standard abbreviation to s.e.d.m. for the standard error of difference between means, we can use this to create a hopefully helpful mnemonic for you. If you substitute the words "estimated variance" for "e" in s.e.d.m, you can remember the calculation as **S**quare root of the "**E**stimated variance" after **D**oubling and **M**ean-ing (dividing by n of the mean).

The importance of the standard error of differences between means

At the beginning of this chapter, I identified the standard **error** of differences between means (s.e.d.m.) as the "likelihood yardstick" for judging the importance we can attach to the size of a difference between two means. It is the yardstick we use most often in biology, since differences between means are the main results of our experiments.

Differences of up to 1 s.e.d.m. between the means of two samples would arise quite frequently by chance (68% of the time, see page 28) when taking many samples from the same population (the two means are just different estimates of a single true mean). However, differences exceeding 2 s.e.d.m. would occur by chance less than 1 in 20 times (5% chance). This would put them in the "unlikely" category, meaning it is unlikely that two samples with those means could be drawn from the same population of numbers. In other words, we would conclude that it is much more likely that the two samples had come from two populations with **different** true means.

That is the basis of many statistical significance tests – "is the probability of these two means being drawn from the same population less than 5%." If so (look back to page 43), we would declare the difference as "statistically significant" at $P < 0.05$ (P stands for "probability," and 0.05 is how we express 5% in probability terms where 1 is certainty), or that the two means are "significantly different" ($P < 0.05$).

You should therefore make sure you understand (i) what the s.e.d.m. actually measures and why it is central to testing the results of experiments and (ii) how to calculate it from the variance (s^2) shown by the individual observations (see the next section below) before tackling the next chapter. There, we begin to put the s.e.d.m. through its paces.

Summary of this chapter

This chapter is so important, that it's worth recapping what it contains.

Box 7.1 in this chapter calculated the *standard error of differences between means* ($n = 3$) by actually taking 100 batches of 3 eggs at random, obtaining the means weight of each batch, pairing these means, and obtaining the 50 differences between the means in each pair. These 50 differences provided a data set of which the variance could be calculated directly by the normal algorithm for calculating the variance of a set of numbers. The standard error of differences between means was then the square root of this variance.

This calculation involved a sample of 300 eggs to obtain 50 differences between pairs of means.

In biological experiments, we usually only have one estimate of the mean for each experimental treatment, and thus only one difference between means.

One can't calculate the variability of one number – it hasn't got any! So **the important message of this chapter** is that we can get the variance of differences between means from the variance shown by the population of single observations – by **doubling** this single-observation variance and then **dividing by the *n*** of the mean (i.e. the number of observations contributing to the mean).

Figure 7.4 illustrates the contents of this chapter, using gray circles for means and solid black circles for the individual numbers from which each mean is calculated. Figure 7.5 then shows, in schematic form, the process of reaching the s.e.d.m. from the variance calculated from the individual numbers (observations). Finally, Fig. 7.6 completes Fig. 7.1 by adding, under the standard normal distribution curve, the scales for the *standard deviation of differences* and also the *standard error of differences between means*. The four scales for egg weight in this figure represent the four statistics highlighted in bold type along the bottom line of Fig. 7.5.

Remember that, whatever we are measuring, we expect 95% of the population of numbers forming a normal distribution to fall in the range between 2 standard error worths on either side of the mean. Look at Fig. 7.5. You will see that we ought to be surprised if we picked up just one egg on our egg farm and found it weighed more than 68.3 g, since this is mean + 2 standard deviations. Again from Fig. 7.5 , you will see that we ought to be equally surprised if the total weight of 3 eggs picked at random was over 200 g (mean 66.8 g). Incidentally, 3 eggs weighing only 89 g in total (mean 63 g) would be just as unusual, as would 2 eggs differing in weight by more than 4.6 g, or if we found a difference in mean weight for 2 batches of 3 eggs that was more than 2.6 g. All have less than a 1 in 20 chance (less than 5% or $P < 0.05$) of turning up on the egg farm. All are equal to or more than their respective mean $\pm$ twice their respective standard deviation/error (see Fig. 7.6).

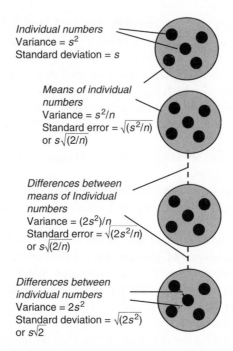

Individual numbers
Variance = s^2
Standard deviation = s

Means of individual numbers
Variance = s^2/n
Standard error = $\sqrt{(s^2/n)}$
or $s\sqrt{(2/n)}$

Differences between means of Individual numbers
Variance = $(2s^2)/n$
Standard error = $\sqrt{(2s^2/n)}$
or $s\sqrt{(2/n)}$

Differences between individual numbers
Variance = $2s^2$
Standard deviation = $\sqrt{(2s^2)}$
or $s\sqrt{2}$

Fig. 7.4   Differences between two means illustrated, with the notation for estimates of variability associated with individual numbers, means, and differences between both. Large gray circles represent means derived from the five individual numbers (small black circles) they contain.

We expect the means of two samples from the same pool of possible numbers forming a normal distribution to be within such limits, i.e. within 2 s.e.d.m. of each other. If the difference is greater than this, then the probability is the samples came from **different** pools of numbers.

It is this last sentence which leads us on to the next chapter, which is our first real use of statistics in an experiment. There's something else important to say at this stage. So far I've talked about the factor of 2 in relation to standard deviations/standard errors as taking us into the "improbable" tail areas of the normal distribution. I've not wanted to keep repeating that it is 2 (to be accurate, actually 1.96!) only when we have a very large number of samples. Since our estimate of the standard deviation/error in small samples has to be suspect (see page 36 and Fig. 6.2), the factor 2 has to be enlarged progressively as samples get smaller. This will be explained more fully in the next chapter.

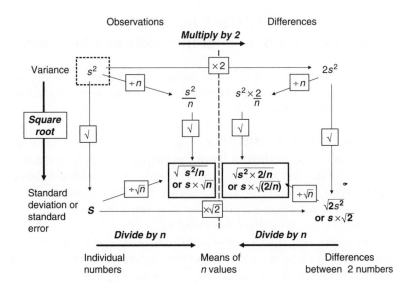

Fig. 7.5 Schematic representation of how the standard error of difference between means (highlighted by rectangles with a solid outline) may be derived from the variance of single observations (highlighted by rectangles with a broken outline).

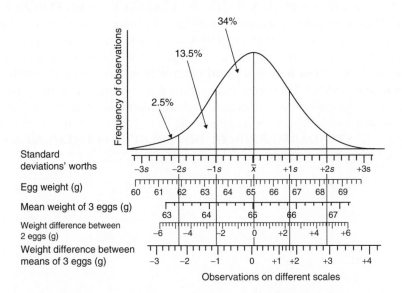

Fig. 7.6 Figure 7.1 of the normal curve for egg weights (showing the proportions of the observations included in various ranges of the standard deviations/errors worth scale) completed by the addition of scales for weight differences between 2 eggs and differences between mean weights of three eggs. Note that all scales have to be at different "magnifications" in order to fit the standard deviations/errors scale.

EXECUTIVE SUMMARY 2
Standard error of a difference between two means

We can calculate variance (s^2) and standard deviation (s) for any set of numbers, but in this book s^2 and s refer throughout to the values calculated from individual numbers:

1 For **INDIVIDUAL NUMBERS** we calculate the variance as the **mean ($n - 1$) squared deviation from the mean**, although on calculators we use the simpler way to get the same answer = **added squares − correction factor)/($n - 1$)**. We square root the variance (s^2) to find the standard deviation (s). These are, as pointed out above, the s^2 and s referred to in the other expressions below.

2 For numbers which represent the **MEAN** of several (n) numbers, the variance of such means is s^2/n; standard deviation (for means it is called STANDARD ERROR) is therefore $\sqrt{s^2/n}$ which can also be written as $s/\sqrt{n}$.

3 For numbers which represent the **DIFFERENCES BETWEEN TWO INDIVIDUAL NUMBERS** taken at random, variance is $2s^2$; standard deviation is therefore $\sqrt{2s^2}$ which can also be written as $s\sqrt{2}$.

The combination of (2) and (3) is the **VARIANCE OF DIFFERENCES BETWEEN MEANS** of several (n) individual numbers. Rather than differences between one number and another, we are measuring the variance of differences between mean values, i.e. the variance of **lots of data** for the **mean of one set of n numbers subtracted from the mean of another set of n numbers.** The appropriate variance, assuming that the variances of the two sets of numbers are equal, is the combination of doubling s^2 (for differences) and dividing by n (for means) – i.e. $2s^2/n$. The square root of this is the **STANDARD ERROR OF DIFFERENCES BETWEEN MEANS (s.e.d.m.)** – $\sqrt{2s^2/n}$ which can also be written as $s\sqrt{2/n}$.

Spare-time activities

At this stage, accept any difference as "statistically significant" which is at least twice as large as the appropriate standard deviation/error.

1 Mean yields of 10 strawberry plants in a uniformity trial were (in grams): 239, 176, 235, 217, 234, 216, 318, 190, 181, and 225.
   (a) What is the variance of the yield of individual strawberry plants?
   (b) On the opposite side of the tractor path, the mean yield of 10 plants of the same variety was 240 g. Is there any evidence that the two areas should not be combined for future experiments?

2 Seven observers were shown, for five seconds per dish, five dishes containing mustard seeds. The observers were asked to estimate the number in each dish, unaware that there were exactly 161 seeds in every dish! The mean guesses per observer across all five dishes were: 183.2, 149.0, 154.0, 167.2, 187.2, 158.0, and 143.0.
   What is the minimum increased number of seeds that you might expect a new observer to be able to discriminate from 161 seeds in five guesses?

3 Sampling of a field of leeks shows that the mean weight of 100 individual plants is 135 g with a variance of 37.4.
   What weight in grams can you attach to the following?:
   (a) A leek 2 standard deviations greater than the mean.
   (b) A mean weight (of 5 leeks) of minus 0.7 standard errors.
   (c) A weight difference between two individual leeks of plus 1.4 standard deviations.
   (d) The maximum mean weight of eight leeks significantly lower than the mean weight of the initial 100 leeks sampled.

8

The *t*-test

Chapter features

Introduction

The *t*-test is perhaps the most important statistical test used in relation to biological variation. In the form described in this chapter, it is a complete test on original data in its own right. However, since it compares means with the *standard error of differences between means*, it is also used in the later stages of other statistical manipulations of original data (see Chapter 10).

The *t*-test enables us to determine the probability that two sample means could have been drawn from the same population of numbers, i.e. the two sample means are really two estimates of one and the same mean. Only if that probability is sufficiently low (most biologists use the 5% probability level, see Chapter 7) do we claim a "statistically significant difference" between the two means, and that it is therefore unlikely that the two means could have been sampled from a single population of numbers (Box 8.1), and therefore are much more likely to be estimates of means which truly differ in magnitude.

The principle of the *t*-test

The *t*-test asks: "Is the difference between the two means large enough, in relation to the variation in the biological material, that it is reasonable to conclude that we have estimates of two truly different means?" – or to put it in the "negative" – "is it likely that the difference between the two means is just a bad estimate of a zero difference?"

BOX 8.1

Ten plants in a plot of lettuces, all from the same seed packet (variety), might weigh 500, 450, 475, 435, 525, 430, 464, 513, 498, and 443 g. If we had sampled only five lettuces, the mean of the first five weights would be 477.0 g. Had we sampled only the last five, the mean would have been only 469.6 g, a difference of 7.4 g. Now in this case we know that both means are estimates of the same figure, the true mean of the whole plot of lettuces – somewhere around 473 g. The difference of 7.4 g above has arisen by chance. But what would be our view if the two means had come from two different lettuce varieties? Could we really claim that the first variety was the better yielder? I don't think so, not without an objective statistical test.

Hopefully it reminds you of the last chapter if I point out that we normally, as in the lettuce example, only have one mean from each variety. Our "yardstick" for evaluating the one difference between means that we have is the *standard error of such differences between means*. Fortunately we can derive this s.e.d.m. from the variability of individual lettuces in the sample (from their variance: doubling it and dividing by the number in the sample, see Chapter 7).

Again a reminder from the previous chapters: *We normally accept that a 1 in 20 or less chance (P = or <0.05 where P = 1 is 100% certainty) of drawing the means from the same pool of numbers constitutes a "significant" difference between those means.*

This level of probability is of course arbitrary. Even double the probability (the 10% chance, $P = 0.1$) is clearly a result of interest; we would be arrogant to claim that we had established there is **no** true difference between the means. Equally we should not take $P = 0.05$ as a sudden "cut-off." We must always remember, when claiming a statistically significant result at this level of probability, that there is a 1 in 20 chance we have drawn the wrong conclusion!

The *t*-test in statistical terms

- We erect the hypothesis (the so-called "null" hypothesis) that there is no true difference between the two means, i.e. the difference found is a bad estimate of a true zero difference.
- We use the individual sampled numbers contributing to both means to obtain a combined (pooled) estimate of the background variability (as variance) of the biological material.

- By three simple steps, we convert this pooled estimate of variance to the *standard error of differences between means* (or *s.e.d.m*) of *n* numbers drawn from the same population.
- Given that the true average difference of means drawn from the same population is zero, we assess how many "standard errors worths" (i.e. *s.e.d.m.* worths) away from zero the difference between the two means represents.
- If this yardstick takes us into a range which makes it unlikely (less than a 1 in 20 chance) that the difference between the means is a bad estimate of zero, we conclude we have disproved the null hypothesis we began with, i.e. the means really are different.

Why *t*?

So the *t*-test boils down to:

$$\text{Is the } \frac{\textit{difference between the means}}{\textit{s.e.d.m.}} \text{ big enough?}$$

What is "big enough?" For very large samples (see earlier chapters, particularly Chapter 7) of perhaps more than 60 numbers each, "big enough" means >2 (actually >1.96), i.e. >1.96 s.e.d.m. This factor of 1.96 is because, in the theoretical normal distribution, numbers more than 1.96 standard deviations away from their mean occur with a frequency of less than 5 %. These limits of mean ± 1.96 standard deviations/errors worths are known as the "95% confidence limits."

Now go back to page 53 and to the paragraph which begins "Remember that, whatever we are measuring ...." This paragraph really already describes the *t*-test in principle. Can you see this?

Of course, we can set any per cent confidence limits other than 95% that we like in terms of standard deviations/errors worths. As pointed out at the end of Chapter 7, we need to amplify this factor of 1.96 for our 95% confidence limits as our samples get smaller to compensate for the increasingly poor estimate of the true variance (Box 8.2 and Fig. 8.1). It is obvious from Fig. 8.1, and also intuitively so, that our ability to measure true variance with any accuracy decreases as true variance increases and/or sample size decreases. Underestimating true variance could lead us to use too small a standard error for differences between means, and lead us to claim a significant difference between two means when the null hypothesis of no difference is actually true (the jargon for this is a "Type I error"). Over-estimated variances could lead to the opposite – we fail to reject the null hypothesis when we should ("Type II error").

> **BOX 8.2**
>
> To illustrate how good or how bad are estimates, based on small samples, of the true variance of a population of numbers, I generated four theoretical normal distributions of different variability around the same mean of 115.
>
> The variances of these four distributions were very different at 460, 230, 115, and 57.5. I then took six random samples from each distribution, at each of six different sample sizes: 5, 10, 20, 40, 80, and 160 numbers. The variances of these 36 different samples are plotted against sample size in Fig. 8.1, with the true variance shown by the broken horizontal line.

Tables of the *t*-distribution

These amplification factors for different size samples (as measured by degrees of freedom, i.e. $n - 1$) can be found in tables of "t" values (Appendix A2.1). They were compiled early in the 20th century by the statistician William Gossett, using the pen-name of "Student" – the amplification factor is thus known as "Student's t." The principal of Student's t-distribution is that it widens the confidence limits in terms of standard deviations/errors worths as samples get smaller. These limits widen to the degree that the per cent of the possible numbers/estimates included theoretically remains constant regardless of sample size.

The requirement that only 5% of differences between means should fall outside the boundary is approached by multiplying the standard deviation/error by a different t for $P = 0.05$ in the table (Appendix A2.1) as sample size (measured as $n - 1$) changes. To see this in action, go to Appendix A2.1, and locate the value of 1.960 at the bottom of the column headed 0.05 (i.e. $P = 0.05$). Note that the $n - 1$ value (in the extreme left column) applicable to the t of 1.96 is very large – infinity, in fact! Now run your finger up the column and note how t increases until it reaches the huge value of 12.7 at the top of the column, where only two samples are involved ($n - 1 = 1$).

How t increases as sample size decreases is also illustrated by Fig. 8.2 by the widening of the shaded areas from right to left (see Box 8.3 for explanation). Widening the *s.e.d.m.* × t limits for small samples has countered the danger of underestimating the true variance in Fig. 8.1 and thus the danger of Type I error (see above). However, there is always a balance, and overestimation of s.e.d.m. (leading to the danger of Type II errors) applies to more of the data sets.

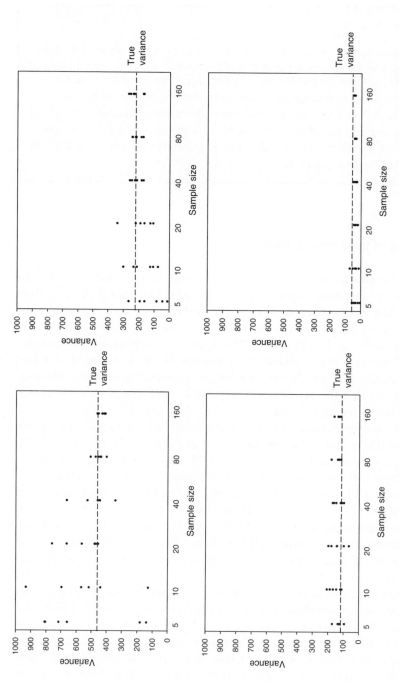

Fig. 8.1 Improved estimate of true variance (dotted horizontal lines) with increasing sample size in four populations with different variability. Variance = top left, 460; top right, 230; bottom left, 115; bottom right, 57.5.

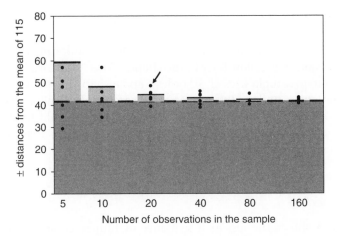

Fig. 8.2 95% confidence limits for observations of the most variable distribution in Fig. 8.1 (i.e. $s^2 = 460$). Three confidence limit calculations are shown. The dotted line shows the true spread on either side of the mean encompassing 95% of observations (i.e. true $s \times 1.96$). The solid lines bounding the lighter gray columns for each sample size show the confidence limit after multiplying the true s by the appropriate tabulated *t*-value for the relevant sample size, and the data points show the 36 points of Fig. 8.1 converted to confidence limits by multiplying the observed s for each point by 1.96.

BOX 8.3

Figure 8.2 uses the variance data of the most variable distribution from Fig. 8.1 (where variance $= 460$) to calculate 95% confidence limits, i.e. (the standard deviations' worths which encompass 95% of the observations).

Confidence limits of the sampled data: Each of the 36 sample variances (s^2) from the top left-hand data set in Fig. 8.1 has been square rooted to calculate the standard deviation (s) and this has then been multiplied by 1.96 (*t* which ignores the small size of the samples).

The true confidence limits of the distribuition from which all the samples were taken: This is of course the same for all sample sizes as $\sqrt{460} \times 1.96$, and is shown by the broken horizontal line.

Confidence limits of the sampled data magnified by t for the relevant sample size: This is obtained by multiplying s by t for $P = 0.05$ at $n - 1$ (for the different sample sizes) degrees of freedom. It is shown by the solid line bounding the gray area of sampled variances lying within the confidence limits for that sample size.

You may be underwhelmed by the number of points ouside these $s \times t$ confidence limits, but bear in mind that there is very little of the distribution beyond these limits. Thus, ignoring the one extreme outlier for samples of 10 observations, the next most aberrant point (arrowed) only represents a further 0.1 $s \times t$ worths beyond the solid line for its sample size (approximately a 96% confidence limit)

The standard *t*-test

I refer to this as the "standard" test, as it is probably the one most used; alternative versions will follow later in the chapter. This "standard" test is most appropriate to sample sizes of 30 or less, and also to larger samples where the variation around the two means being compared is similar. Comparing a mean of varying numbers with a series of zeros would be stretching the test well beyond its valid limits (see "*t*-test for means associated with unequal variances" below).

The procedure

Figure 8.3 shows the test as a sequence of numbered stages. You may, however, prefer the version in Box 8.4, where the test is illustrated on a set of data. Either way, the text which follows picks up on what each stage is doing, following the white numbers on black circles in Fig. 8.3 and Box 8.4. The typical *t*-test evaluates the significance of the difference between two means, each of a number of individual observations. Hence the two columns (series *x* and *y*) of individual observations in Box 8.4. Often the two means would be based on an equal number of observations, but I have used unequal numbers in Box 8.4 to show that the standard *t*-test can still be used.

❶*Some summary numbers*

Initially, we compile the simplest of statistics for each column of observations (*x* and *y*), namely (using the *x* series as the example) their number (n), their total ($\sum x$), and the mean ($\sum x/n = \bar{x}$).

Pooled variance

This computation is the main new concept the *t*-test introduces. Combining the variances of the *x* and *y* series by adding them together just wouldn't give the right answer. Once we have divided by $n - 1$, variability is no longer additive (Box 8.5). But one *can* add sums of squares together. If you go back to what sums of squares really are (page 15), they are the sums of numbers, each of which is a squared deviation from the mean. Each of these squared deviations is added to the ones before – whether you add them in groups of one, two, three, or four won't change the final total (Box 8.5). Grasp this concept – you will find it essential for understanding the principle of the analysis of variance (Chapter 10).

SUMMARY OF *t*-TEST

Numbers represent suggested order of calculation

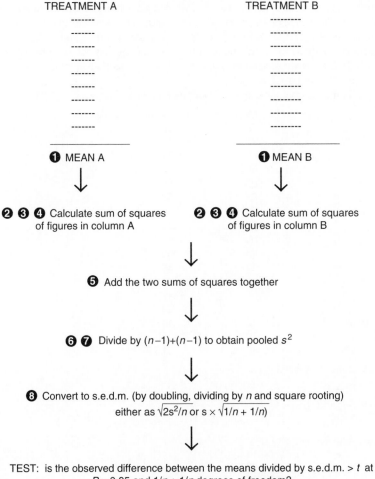

Fig. 8.3 Summary of the procedure involved in the standard *t*-test.

❷So we calculate the correction factor (for sums of squares by the quick method) separately for series x and series y. ❸Similarly we add the squared numbers separately for x and y and so, after subtracting the appropriate correction factor for each series, ❹we arrive at a separate figure for the sum of squares for each series. Note that we have used the two different series totals in the correction factors for the two series. This is equivalent to using

BOX 8.4

We have a sample of 12 men and 14 women who have been subjected to the horror of a statistics examination and want to test whether there is any evidence that the difference of 2.71% in the mean per cent mark awarded (59.08 for men as against 61.79 for women) is statistically significant. Examination marks, especially those between 80% and 20%, can be expected to be normally distributed, so transformation (which is often necessary with percentages, see page 40) is not necessary.

The marks and subsequent calculations are as follows:

	Men (x series)	Women (y series)	
	75	80	
	72	76	
	68	74	
	66	70	
	65	68	
	65	68	
	60	66	
	58	65	
	50	62	
	48	58	
	42	56	
	40	43	
		40	
		39	

❶ n   12   14

Total $\sum x$ 709   865 $\sum y$

Mean $\overline{x}$ 59.08   61.79 $\overline{y}$

❷ Correction factor
$(\sum x)^2/n$ 41,980.08   53,444.64 $(\sum y)^2/n$

❸ Added squares
$\sum x^2$ 43,460.99   55,695.00 $\sum y^2$

❹ Sum of squares
$\sum x^2 - (\sum x)^2/n$ 1480.91   2250.36 $\sum y^2 - (\sum y)^2/n$

❺ Pooled sum of squares
$= \sum x^2 - (\sum x)^2/n + \sum x^2 - (\sum x)^2/n$ 3731.26

❺ Pooled degrees of freedom
$= 11 + 13$ 24

❻ Pooled variance   155.47

❼ Pooled standard deviation   12.47

❽ s.e.d.m. $= s\sqrt{1/n + 1/n}$
$= 12.47\sqrt{1/12 + 1/14}$ 4.91

❾ $t = \dfrac{\overline{x} \text{ and } \overline{y} \text{ difference}}{\text{s.e.d.m.}} = \dfrac{59.08 \text{ and } 61.79 \text{ difference}}{4.91} = 0.55$

BOX 8.5

To illustrate pooling variance, let's just take the results for men (x series) from Box 8.4, and compute the sums of squares by actually using deviations from the mean (which is 59.08) rather than the quicker *added squares – correction factor* method we have got used to by now:

x	$(x - \bar{x})$	$(x - \bar{x})^2$	$\sum(x - \bar{x})^2$
75	15.92	253.45	
72	12.92	166.93	
68	8.92	79.56	
66	6.92	47.88	547.82
65	5.92	35.05	
65	5.92	35.05	
60	0.92	0.85	
58	−1.08	1.17	
50	−9.08	82.45	
48	−11.08	122.76	
42	−17.08	291.72	
40	−19.08	364.04	933.09
		1480.99	1480.91

Not surprisingly, whether we sum the squared deviations one after the other, or in two blocks above and below the line (547.82 + 933.09), we get the same answer (1480.91) and the variance is $1480.91/n-1 = 1480.91/11 = 134.63$.

But what happens if we calculate the two variances in the right-hand column and then add them together?

Above the line variance is $547.82/3 = 182.61$, and below the line $933.09/7 = 133.30$. This gives a combined variance of 315.91, much larger than the right answer above. Yes I know the divisors were different, 11 as compared with $3 + 7$, but that clearly can't explain the huge discrepancy between 134.63 and 315.91. No, we have to pool the sums of squares BEFORE we divide by $n - 1$.

It makes total sense! The components (182.61 and 133.30) of the larger answer (315.91) are really estimates, from two small samples, of the variance of 134.63 estimated from all 12 samples. So we should expect an answer almost double that 134.63.

two different means if we were to calculate the sums of squares directly by adding squared deviations from the mean. Using the two different totals removes from the pooled variance the variability introduced by having two different treatments, and so we only measure the variation of the biological material in each column around its respective mean.

❺The pooled sum of squares is no more complex than the sums of squares of the two series added together.

❻The pooled degrees of freedom are perhaps less intuitive. However, if you remind yourself from page 16 that using the sample mean to calculate sum of squares loses a degree of freedom, and that we have used both means in calculating our pooled sum of squares, then we also need to pool two degrees of freedom as $(n-1) + (n-1)$ for the two sums of squares being pooled. So in the example in Box 8.4, with unequal numbers in the two columns, we pool $11 + 13 = 24$.

❼ Remember, variance is *sum of squares* divided by *degrees of freedom* (page 16), so pooled variance *is pooled sum of squares* divided by *pooled degrees of freedom.*

Standard error of differences between means (s.e.d.m.)

This calculation begins with the pooled variance. Remember the mnemonic on page 52. The s.e.d.m. is "the **S**quare root of the **E**stimated variance, after **D**oubling and **M**eaning." In notation this is:

$$\sqrt{\frac{2s^2}{n}}.$$

This is easy if n is the same for both means: we double the pooled variance, divide by the n shared by both means, and then square root. ❽However if, as in our example in Box 8.4, the number of observations for each mean differs, we need to get the pooled standard error (s) obtained by square rooting the pooled variance. We then develop our calculation for the s.e.d.m. from the algebraic identity that

$$\sqrt{\frac{2s^2}{n}} \text{ can also be written as } s\sqrt{\frac{2}{n}}.$$

Hopefully your algebra is still up to this? Although either of these identities is fine if the two means are based on the same number of observations, the second one enables us to go one step further where unequal numbers of observations are involved. The $2/n$ can be rewritten as $1/n + 1/n$, i.e. the s.e.d.m. can be rewritten as $s\sqrt{1/n + 1/n}$, giving us the opportunity to insert the two different n values of 12 and 14 (see step ❾ in Box 8.4). Thus we obtain an s.e.d.m. for the statistics marks of 4.91.

❾ *The actual t-test*

To recap from the start of this chapter, the *t*-test asks:

is the $\dfrac{\textit{difference between the means}}{s.e.d.m.}$ big enough to be statistically significant?

where big enough to be statistically significant is normally taken as the tabulated value of t at $P = 0.05$ for the degrees of freedom of the pooled variance $(11 + 13 = 24)$, i.e.:

is the $\dfrac{\textit{difference between the means}}{s.e.d.m.} > t$

So for the mean statistics marks of men and women in our Box 8.4 example, the *t*-test is:

$$\frac{59.08 - 61.79}{4.91} = 0.55$$

You may think it should be –0.55? Well, the sign is immaterial. 0.55 is the **difference** between the means; the two means could equally have been reversed in order. Remember that, but in any case in future we'll express a difference like $\bar{x} - \bar{y}$ where we use the difference as an absolute value (i.e. regardless of direction) as "$\bar{x}$ **and** y **difference**." The value of t in the tables at $P = 0.05$ and 24 degrees of freedom is 2.06 (Appendix A2.1). The difference between the means is only half the s.e.d.m. and nowhere near the 95% confidence limit for differences between statistics mark means that a t of 2.06 represents. The difference of about 3 marks is therefore the sort of difference between two means that would often occur if we sampled the marks of either gender twice. We have not disproved the "null hypothesis" (page 59). The difference between the means is negligible in terms of the variation that occurs between individual students and we have absolutely no justification for suggesting there is any difference between the statistics marks of our men and women students. The t value would have had to reach or exceed 2.06 before we could claim any difference between the mean marks for the sexes.

t-test for means associated with unequal variances

The "standard *t*-test" assumes that the two means came from normal distributions with similar variances. In effect we are testing whether the two means could have been sampled from the same population of numbers. Then, particularly useful with small samples, the numerators of the variance formula (i.e. the sum of squares of deviations from the mean) could be combined as a pooled estimate of the variability of the biological material.

Sometimes, however, we know we are dealing with two distinct populations of numbers, since the variance of the observations for each mean are noticeably unequal (for what "noticeably unequal" means, see the F-test in the next chapter). We may still want to determine whether the means differ – for example, one apple variety may have a clear advantage over another in terms of uniformity of fruit size, but we may nevertheless also want to know that its yield per tree is not inferior. We cannot however pretend that the biological material shows uniform and thus "poolable" variation! However, we do need enough observations around both means to be able to detect this (remind yourself of Fig. 8.1), so the form of the t-test discussed here tends to involve samples in excess of 30.

The s.e.d.m. when variances are unequal

The standard error for difference between means (s.e.d.m.) is most simply remembered as "the **S**quare root of the **E**stimated variance, then **D**oubled and **M**ean-ed" – in notation this is $\sqrt{s^2 \times 2/n}$. The formula is however more usually written as $\sqrt{2s^2/n}$, and the algebra can be tweaked further. Although $\sqrt{2s^2/n}$ is fine for how we use the pooled variance in the standard t-test with equal numbers of observations for the two means, now we don't have a common s^2 and the two means may not be based on equal numbers. However, the notation $\sqrt{s^2 \times 2/n}$ involves $2s^2$, obtained by doubling one (pooled) variance in the standard test. With the t-test for unequal variances we have two variances, each of which we can use once to get $2 \times s^2$, instead of doubling one variance as in the standard test. Thus the s.e.d.m $\sqrt{s^2 \times 2/n}$ can be written as $\sqrt{\mathbf{s^2/n} + \mathit{s^2/n}}$, enabling us to insert both our unequal variances and unequal numbers! (The two s^2/n values in this expression are shown respectively in **bold** and *italics* to indicate they refer to the data for *different* means).

So the *t-test for means associated with unequal variances* is still:

$$\text{is the } \frac{\textit{difference between the means}}{\text{``s.e.d.m.''}} > t$$

but with the s.e.d.m. in quotation marks because it is modified to take account of the fact that there are two variances and not one pooled variance. So the *t*-test becomes:

$$\text{is the } \frac{\textit{difference between the means}}{\sqrt{\dfrac{\mathbf{s^2}}{\mathbf{n}} + \dfrac{s^2}{n}}} > \text{``}t\text{''}$$

where bold and italics refer to the different data for the two means. This can alternatively be written as:

$$\text{is the } \frac{\textit{difference between the means}}{\sqrt{(\mathbf{s^2}/\mathbf{n} + s^2/n)}} > t$$

which is the way of expressing division procedures occurring within a divisor that I tend to use throughout this book.

We again use the $P = 0.05$ column in the t table, but we need to spend a bit of time thinking of what are the appropriate degrees of freedom. With a pooled variance, remember, we used $(\mathbf{n-1}) + (n-1)$ degrees of freedom because the variation represented by the variance was assumed to be that of a pool of $\mathbf{n} + n$ observations. In other words, it is as if we had calculated just one variance from a sample size of $\mathbf{n} + n$. With unequal variances, each is calculated from the rather smaller sample size of just one "n" – and t has to express the poorer reliability we can place on estimates from small samples (Box 8.2 and Fig. 8.1), even though we have two of them.

The magnitude of the s.e.d.m. will be dominated by the variability of the larger set of numbers (Box 8.6), and the difference between two means is less likely to be significant as the difference between the variance and/or number of observations in the two treatments increases. The solution is

---

BOX 8.6

To show the effect of a differing n in the *t*-test for means associated with different variances, we'll do the t calculation for two different populations (**mean 50 and variance 200** and *mean 25 and variance 50*) three times. In each case the two series will be distinguished in **bold** and *italics*, and the **total** number of observations will be kept constant at 60.

1 Both variances based on 30 numbers:

$$t = \frac{\mathbf{50} \text{ and } \textit{25} \text{ difference}}{\sqrt{\mathbf{200}/\mathbf{30} + \textit{50}/\textit{30}}} = \frac{25}{\sqrt{6.67 + 1.66}} = \frac{25}{2.89} = 8.65$$

(Continued)

BOX 8.6   Continued

2 Larger variance based on larger n (e.g. numbers are **50** and *10* to still add to 60):

$$t = \frac{\textbf{50} \text{ and } \textit{25} \text{ difference}}{\sqrt{\textbf{200/50} + \textit{50/10}}} = \frac{25}{\sqrt{4+5}} = \frac{25}{3} = 8.33$$

Note that *t* is reduced, since the poorer estimate of variability from only 10 samples has raised its s^2/n from 1.66 to 5, with relatively little change to the s^2/n for the sample of 50 observations.

3 Larger variance based on smaller n (e.g. numbers are **10** and *50* to still add to 60):

$$t = \frac{\textbf{50} \text{ and } \textit{25} \text{ difference}}{\sqrt{\textbf{200/10} + \textit{50/50}}} = \frac{25}{\sqrt{20+1}} = \frac{25}{4.58} = 5.46$$

Note *t* is now much reduced because most of the variability comes from the smaller sample to give a huge s^2/n of 20.

The calculations (see page 73) below (using the example in this box and **bold** and *italic* fonts as before) show how d.f. (and therefore the likelihood of a significant *t*) decrease as the numbers of observations in the two treatments (total kept at 60) differ increasingly in either direction from equal *n*. High numbers of observations in the more variable treatment (in **bold**) reduce the d.f. especially strongly. Note that d.f. have been rounded to integer values:

n	*n*	s^2	s^2	d.f.
5	55	200	50	6
10	50	200	50	26
15	45	200	50	49
20	40	200	50	58
25	35	200	50	53
30	30	200	50	43
35	25	200	50	33
40	20	200	50	24
45	15	200	50	16
50	10	200	50	10
55	5	200	50	4
30	30	200	200	58

The last line in the table shows that the calculation gives d.f. of $(\textbf{n}-\textbf{1}) + (\textit{n}-\textit{1})$ if variances and number of observations are both equal.

to reduce the degrees of freedom from $(\mathbf{n-1}) + (n-1)$ against which to look up t in the t table. Box 8.6 therefore gives a calculation which deals with this problem not by recalculating t, but recalculating the degrees of freedom against which to look up t (calculated by the standard method) in the table of t. This of course reduces the significance of any value of t. The calculation is complex, and I fear my understanding does not stretch to deriving it for you logically. You may at least be comforted by the result (see Box 8.6) that d.f. are $(\mathbf{n-1}) + (n-1)$ if we run the spreadsheet with both variances and n equal for both treatments.

The starting point is on page 45 or Fig. 7.5, where I pointed out that the variance of mean values was the variance of individual observations (s^2) divided by the number of observations in the mean (n), i.e. s^2/n. The variance of the means of the two populations (again distinguished by **bold** and *italic* fonts below) in our t-test is the main factor in the calculation we use for d.f. when these variances are obviously not equal:

$$\text{d.f.} = \frac{(\textbf{variance of mean} + \textit{variance of mean})^2}{(\textbf{variance of mean})^2/\textbf{n} - \textbf{1} + (\textit{variance of mean})^2/n - 1}$$

Box 8.6 shows how the degrees of freedom change as the same number of observations is partitioned differently between two populations of disparate variance.

A worked example of the t-test for means associated with unequal variances

The example chosen includes the complication of unequal numbers as well as unequal variances (Box 8.7) and follows as far as possible the steps for the "standard" t-test from Fig. 8.3 and Box 8.4. Note in Box 8.7, however, that at step ❼ variances for the two series are calculated separately and neither the sums of squares nor the degrees of freedom are pooled as they are in the standard t-test. The combination (not pooling) of the two variances happens at step ❽ in Box 8.7, where each variance (s^2) is divided by its own "n" to convert it to a variance of ***means*** (s^2/n) and then the two are added together to give us the ***doubling*** element required for variances of differences between means $(2 \times s^2/n)$. Square rooting then gives us the standard error of ***differences*** between ***means*** (the s.e.d.m.). ❾ *The actual t-test:* Dividing the difference between the means by the s.e.d.m. gives us a high observed t of 40.86.

The final calculation in Box 8.7 inserts the real numbers into the somewhat complex equation given above for calculating the d.f. appropriate when the two variances are not equal. $(\mathbf{n-1}) + (n-1)$ would be $\mathbf{39} + 28 = 67$, but our calculation to accommodate unequal variances

BOX 8.7

40 tubers of the potato variety "Maris Piper" and 29 of "Jupiter" were obtained as ready-bagged in a supermarket, and weighed individually to test whether the mean tuber weight differed between the varieties, though it was obvious that the size of "Maris Piper" tubers varied very much more than "Jupiter" tubers. As in the text of this chapter, notation for the two populations is distinguished where necessary by using bold text for "Maris Piper" and italic text for "Jupiter." The calculations are as follows:

	"Maris Piper" (x series)	"Jupiter" (y series)	
Columns of individual observations not provided in the cause of brevity			
❶ n	40	29	
Total $\sum x$	4160	2455	$\sum y$
Mean $\bar{x}$	104	85	$\bar{y}$
❷ Correction factor $(\sum x)^2 / n$	432,640.00	207,828.44	$(\sum y)^2 / n$
❸ Added squares $\sum x^2$	432,900.14	207,875.70	$\sum y^2$
❹ $\sum x^2 - (\sum x)^2 / n$	260.14	42.26	$\sum y^2 - (\sum y)^2 / n$
❺ Degrees of freedom	39	28	

❻ Variance (s^2):

$$= \left[\sum x^2 - (\sum x)^2 / n \right] / n - 1 \quad 6.67 \qquad 1.46 \qquad \left[\sum y^2 - (\sum y)^2 / n \right] / n - 1$$

$$s^2/n = 6.67/40 = 0.166 = \quad 1.46/29 = \quad 0.050$$

❼ s.e.d.m. $= \sqrt{s^2/n \text{ for Maris} + s^2/n \text{ for Jupiter}}$

$$= \sqrt{0.166 + 0.050} = \quad 0.465$$

❾ $t = \dfrac{\bar{x} \text{ and } \bar{y} \text{ difference}}{\text{s.e.d.m.}} = \dfrac{104 \text{ and } 85 \text{ difference}}{0.465} = 40.86$

t needed to reach significance at $P = 0.05$:

$$= \dfrac{(s^2/n + s^2 n)^2}{(s^2/n)^2/n - 1 + (s^2/n)^2/n - 1}$$

$$= \dfrac{(0.166 + 0.050)^2}{0.166^2/39 + 0.050^2/28} = \dfrac{0.0467}{0.000707 + 0.0000892*} = 58.65 \text{ (round to 59)}$$

* Many calculators would show this divisor as $7.07E - 4 + 8.92E - 5$ (the E- number showing how many places the decimal point should be moved to the left).

reduces this to 59 (Box 8.7). In the example, this actually makes negligible difference to the outcome. At $P = 0.05$ for 60 d.f. (close to the calculated 59) t is 2.000. Clearly our observed t-value of over 40 is highly significant (t at $P = 0.001$ is still only 3.460!). We can conclude that the mean tuber weight of "Maris Piper" is indeed greater than that of "Jupiter."

The paired *t*-test

This is a very efficient-test for identifying differences between means, especially when the material being studied is very variable. The basic principle of the test is that we calculate the pooled variance, not from the variance of the **individual observations** as done previously, but directly from **differences between pairs of numbers**, differences which are the result of the experimental treatment we are investigating.

A description of this test really has to begin with some suitable data, for example the dry weight of plants grown in two different field soils (Box 8.8).

It is obvious from Box 8.8 that there is high variance within each column stemming from the very different-sized plant species that were used to test the two soils fully. These variances are pretty similar in the two soils and so we could (though we would be foolish to do so) ignore that the data are paired by plant species – and use the standard *t*-test as described earlier:

$$t = \frac{\bar{x} \text{ and } \bar{y} \text{ difference}}{\text{s.e. of difference between means of 10 figures}}$$

As "n" is the same for both soils, we could use the formula:

$$t_{P=0.05,18 \text{ d.f.}} = \frac{\bar{x} \text{ and } \bar{y} \text{ difference}}{\sqrt{2s^2/n}}$$

The pooled variance in the standard *t*-test (remember we add the sums of squares before dividing by degrees of freedom) is:

$$\frac{481.39 + 456.06}{9 + 9} = 52.08 \text{ and } t = \frac{9.07 \text{ and } 8.87 \text{ difference}}{\sqrt{2 \times 52.08/10}} = 0.06.$$

So that's the result of using the standard *t*-test, and we would conclude that the difference between the means of 0.20 is very small in s.e.d.m. terms and is therefore a reasonably close estimate of a true difference between the

---

BOX 8.8

To evaluate the relative merits of two field soils for container-grown plants in a nursery, the growth of one specimen of each of 10 different plant species was measured in each soil. Data recorded were the dry weight in grams of the aerial parts of the plant when ready for sale.

Species	Soil A (x series)	Soil B (y series)	
1	5.8	5.7	
2	12.4	11.9	
3	1.4	1.6	
4	3.9	3.8	
5	24.4	24.0	
6	16.4	15.8	
7	9.2	9.0	
8	10.4	10.2	
9	6.5	6.4	
10	0.3	0.3	
n	10	10	
Total $\sum x$	90.7	88.7	$\sum y$
Mean $\bar{x}$	9.07	8.87	$\bar{y}$
Correction factor $\left(\sum x\right)^2 /n$	822.65	786.77	$\left(\sum y\right)^2 /n$
Added squares $\sum x^2$	1304.03	1242.83	$\sum y^2$
Sum of squares $\sum x^2 - \left(\sum x\right)^2 /n$	481.38	456.06	$\sum y^2 - \left(\sum y\right)^2 /n$
Degrees of freedom	9	9	
Variance (s^2): $= \left[\sum x^2 - \left(\sum x\right)^2 /n\right]/n - 1$	53.49	50.67	$\left[\sum y^2 - \left(\sum y\right)^2 /n\right]/n - 1$

means of zero. We would conclude that the two soils are equally suitable for growing a wide range of plant subjects. Yet if you look at the data in Box 8.8 again, there does seem to be a very consistent pattern of slightly poorer growth in Soil B (except for plant species 10, which attains the same dry weight in both soils and species 3 where growth is a little poorer in soil A). Surely this pretty regularly occurring superiority of Soil A is not just chance variation? The trouble with our standard t-test is that the large variation within the 10 plants in soil A is not paired with the similar obviously variation between plant species in soil B. In other words, the standard t-test takes no notice of the fact that each number in the x series has an obvious y partner. The standard test would give you the same pooled

BOX 8.9

Species	Soil A (x series)	Soil B (y series)	Difference x − y (z series)
1	5.8	5.7	+0.1
2	12.4	11.9	+0.5
3	1.4	1.6	-0.2
4	3.9	3.8	+0.1
5	24.4	24.0	+0.4
6	16.4	15.8	+0.6
7	9.2	9.0	+0.2
8	10.4	10.2	+0.2
9	6.5	6.4	+0.1
10	0.3	0.3	0
Mean	$\bar{x}$ 9.07	8.87 $\bar{y}$	

Total $\sum z$ 2.0

$n - 1$ 9

Correction factor $(\sum z)^2 / n = 4^2/10$ 0.4

Added squares $\sum z^2 = +1^2 + 0.5^2 + (-2^2) + \cdots + 0^2$ 0.92

Sum of squares $\sum z^2 - (\sum z)^2 / n$ 0.52

Variance (s^2) = Sum of squares/$n - 1 = 0.52/9$ 0.058

variance whether or not – or how many times – you shuffled the numbers within each series.

In the paired *t*-test, by contrast, the order of the numbers in each series is crucial – the partners must be kept together as a pair. Then we are able to remove the high variability between the plant species by using as the data the 10 *differences* between the members of each pair (the difference caused by the experimental treatment of a different soil). These differences are shown in Box 8.9.

You will see in Box 8.9 that the 10 differences $(x - y)$ in the individual pairs are numbers (with a + or − sign) which can be used to generate a variance (s^2) of 0.058. As the 10 numbers are differences, their variance is already the *variance of differences between individual observations* (the $2s^2$ of Fig. 6.2 or double the pooled variance in the standard *t*-test). In other words, our calculated 0.058 above is the **e**(estimated variance of individual numbers) and the **d** (doubling for differences) and we then only need the **m** (divide by *n* of the mean) and the **s** (square root) to calculate the s.e.d.m. for the *t*-test.

So our s.e.d.m. is $\sqrt{0.058/10} = 0.076$.

The value of t as usual is the difference between the mean of the x series and that of the y series (i.e. $\bar{x} - \bar{y}$) divided by the s.e.d.m., and so is $0.2/0.076 = 2.262$.

As we have used the variability of 10 numbers (the 10 differences rather than the 20 dry weights) for our variance, degrees of freedom for t are only 9 compared with $18(9 + 9)$ had we used the standard t-test.

The tabulated value for $t_{P = 0.05}$ for 9 d.f. is exactly 2.262. Our calculated t is also 2.262 and so is significant at exactly the 1 in 20 chance of drawing the wrong conclusion. This is an extremely borderline result for rejecting the null hypothesis that our two soil types have no effect on plant dry weight; there is really insufficient evidence either way.

So the comparison of the unpaired and paired t-test of the same data (Box 8.8) is as follows:

The unpaired test gave a t of 0.060 with 18 d.f. Tabulated t ($P = 0.05$) was much higher at 2.101; we accept the null hypothesis and conclude that plant dry weight in both soils is virtually the same.

The paired test gave a t of 2.262 with 9 d.f. Tabulated t ($P = 0.05$) is also 2.262: As pointed out above, this casts considerable doubt on the conclusion from the unpaired test that plant dry weight does not differ between the soils.

The paired t-test (in spite of lower degrees of freedom) is clearly far more sensitive and far more appropriate where data are paired (as in Box 8.8).

Pair when possible

Experiments involving only two treatments and suitable for analysis by the t-test can often be improved by building "pairing" into the design. For example, with experiments where plants are in pots, they can be size-matched in pairs before the two treatments are applied. Other pairing could arise from using two leaves per plant, the right and left halves of the same leaf, two fish in each of several fish tanks, etc. However, pairing does reduce the degrees of freedom for t (which then has to be larger to reach statistical significance). It therefore does pay to look for variability worth pairing in the experiment before deciding whether or not to pair the observations.

In any event, nothing is ever lost by pairing observations when setting up the experiment, if there is an obvious reason for doing so. This is a rare instance in statistics where it is "legal" to decide – after the data have been recorded – whether to analyze those data by one calculation (paired) or a different one (unpaired). If the paired analysis cannot capitalize on the pairing sufficiently to outweigh the higher t required by the loss of degrees of freedom, then an unpaired test is still perfectly valid. But beware! The reverse is not true. You cannot set up an experiment as "unpaired" and then pair the data "after the event," e.g. by making the largest number in each series a pair, and so on. Certainly not!

EXECUTIVE SUMMARY 3
The *t*-test

By this time, it is assumed that you have "come to terms" with the concepts:

1 that we measure variability by "variance";
2 that we can derive variances for numbers which represent means or difference values – all from the variance of individual values;
3 that standard error (for means or differences between means) is the square root of the appropriate variance;
4 that if variance of individual values is s^2, and the variances of two populations is equal, then the standard error appropriate for judging "differences between their means" is $\sqrt{2s^2/n}$;
5 that we normally use the admittedly arbitrary convention that any value which is more than some 2 standard errors different from the mean is regarded as a significantly "unlikely" event (i.e. having less than a 1 in 20 chance of occurring).

The *t*-test is the significance test normally used to test differences between two means in biology. After sampling two experimental treatments, we finish up with two mean values, one for each treatment. The **original data** are the basis for measuring **"sampling variation"** (i.e. variance of the **individual** values). The stages of the test (where variance in the two treatments can be assumed equal) are as follows:

1 We see if it is possible to **reject** the hypothesis that there is no real difference between the means of the two treatments.
2 If this "null" hypothesis were true, then whatever difference we have found between the means (i.e. **the mean A and mean B difference)** is in the range of "likely" differences, i.e. likely to arise by chance sampling when taking small samples from two populations of numbers whose real means are identical (i.e. **for the whole populations, the mean A and mean B difference = 0**).
3 To test the null hypothesis, we need to know what *size differences* between *two **identical** means* could reasonably be expected just as a result of sampling. We can then assess whether the observed difference between the treatments is large enough for it to be **unlikely** that sampling could account for the *difference* between the two means that we have observed.
4 We calculate the **sum of squares** (the top of the variance expression) separately for each column of figures. This (remember?, although

we calculate it by *added squares* minus the *correction factor*) is really the summed squared deviations of the numbers from their mean, and so measures variation around the mean (we have eliminated any bias for different means in the two columns by theoretically subtracting the mean of its column from each number when we use the column total in the correction factor). We calculate the combined (= **pooled**) variance by adding the two *sums of squares* (i.e. the two **added squares minus correction factors**) and then dividing by the sum of the two degrees of freedom [**(n − 1** for one population) + **(n − 1** for the other population)]. This is the pooled variance (s^2) for both columns of numbers.

5 **If the number of observations (n) in the two columns is the same**, we then double s^2 (to measure the variance of differences) and then divide by n (to measure variance of means) to reach **$2s^2/n$**, the variance of the differences we would expect sets of two means of n numbers to show (because of sampling variation) where there is no real treatment difference. The square root of this is the standard error of expected differences between two means on the null hypothesis.

6 **If the number of observations in the two columns differs**, we first square root the pooled variance to get the pooled standard deviation. We now have to multiply this by $\sqrt{2/n}$ to get the equivalent of $\sqrt{2s^2/n}$ above, and this we do by multiplying s by $\sqrt{1/n + 1/n}$, where the two n's are the two different n's for the two columns.

7 Either way, we have calculated the standard error for the sizes of differences we would expect to occur between two means (by chance sampling from two populations) which actually have the **same** mean (null hypothesis). This standard error (**standard error of differences between means = s.e.d.m.**) is the yardstick for judging how **in**accurately it is possible to measure a true difference of zero between the means, the situation under the null hypothesis. In other words, we know that the **differences between sets of two mean values** taken where there is **no true difference between means** will have a mean of zero and a standard error represented by the s.e.d.m. An observed difference greater than twice the s.e.d.m. is taken to indicate that the null hypothesis is untrue, i.e. the observed difference between the means is significantly large.

8 The test is therefore: is the observed difference more than 2 × s.e.d.m., i.e. is the **observed difference divided by s.e.d.m. > 2?**

"2"(actually 1.96) is the criterion for very large samples. For smaller samples we have to increase the "2" somewhat as given in "t" tables. Here we look up the column for the 1 in 20 risk ($P = 0.05$) of drawing the wrong conclusion against the row for $[(n-1) + (n-1)]$ degrees of freedom. So the test finishes up as: **is the observed difference/s.e.d.m. > *t* ?**

Variations of the test are suitable for situations where variance in the two treatments clearly differs, or where pairing between samples from the two treatments is possible. Consult Chapter 8 for the procedures involved.

Spare-time activities

1 The yields (kg per plot) of two varieties of beans were:

Variety A – 3.4, 4.2, 4.8, 3.7, 4.2, 4.3, 3.3, 3.6, 4.2, 3.1, 4.8, 3.7, 4.7, 3.9, 4.8, 4.0, 3.8, 4.9, 4.5

Variety B – 4.6, 4.5, 3.8, 4.0, 4.7, 4.9, 4.3, 5.0, 3.8, 4.1, 5.1, 5.2, 4,9, 4.3, 4.2, 3.9, 4.4, 4.9, 4.6

Does either variety significantly outyield the other ? If so, which variety and by how much?

2 The number of seeds (out of 50) of *Tagetes* germinating under two environmental conditions was:

Bought-in compost 36, 43, 44, 39, 44, 50, 39, 44, 46,

39, 42, 50, 42, 46, 39, 38, 49, 38

Own sterilized soil 28, 31, 25, 28, 28, 27, 32, 24, 33, 33

Has the grower done enough tests with his own soil to be able to decide which potting medium to use in future ? (ignoring any difference in cost).

3 The distances travelled (cm) by two green chlorophyll extracts applied as adjacent spots on 13 paper chromatograms were as follows:

Sample	Extract A	Extract B	Sample	Extract A	Extract B	Sample	Extract A	Extract B
1	5.8	4.0	6	6.5	5.1	11	5.1	3.8
2	6.6	6.1	7	5.0	5.2	12	5.6	4.3
3	7.3	4.5	8	4.9	5.2	13	6.2	5.7
4	6.3	4.9	9	5.6	5.4			
5	5.9	5.2	10	5.7	5.6			

These data are "paired" runs, since such are generally of different lengths.

Has the experimenter statistical grounds for confidence that the two green constituents of the two extracts represent different compounds?

9

One tail or two?

Chapter features

Introduction

Significance tests can be "two-tailed" or "one-tailed." The difference is crucial in attaching the correct probabilities to our conclusions from an experiment, yet many biologists do not think about the distinction at all. The t-tests described in Chapter 8 all involved two-tailed tests, and I know colleagues who claim never to have used a one-tailed test. I don't like to tell them that they have, every time they have done an analysis of variance! You will have to wait until the next chapter before we get onto "Analysis of Variance" in any detail, but let me tell you now that it ends in comparing the variance caused by the intentional treatments designed into the experiment with the unavoidable background variance of the biological material. We divide the treatment variance by that of the background to give the statistic F, otherwise very appropriately known as the *variance ratio*.

Why is the analysis of variance *F*-test one-tailed?

In the analysis of variance, we are testing whether the treatment variance is "significantly" **more important** compared to background variance (i.e. is it significantly larger?). Is it outside the critical limit on the PLUS side which embraces 95% of the variation in background variances that might be sampled by chance? If it is within this limit, including the whole of the MINUS side (i.e. F is less than 1), then treatments have had no noticeable effect. We attach no importance to even an exceptionally small F value right at the extreme of the minus side – it just means the treatment effect is exceptionally negligible!

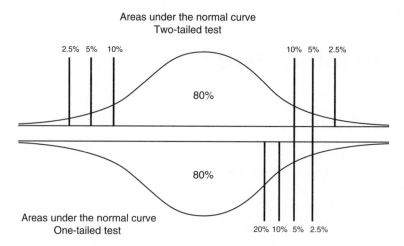

Areas under the normal curve
Two-tailed test

2.5% 5% 10%

10% 5% 2.5%

80%

80%

Areas under the normal curve
One-tailed test

20% 10% 5% 2.5%

Fig. 9.1   Areas under the normal curve appropriate to probabilities of 20% ($P = 0.2$), 10% ($P = 0.1$), 5% ($P = 0.05$), and 2.5% ($P = 0.025$) in a two-tailed (upper curve) and a one-tailed (lower curve) test. In the upper curve, the per cent figures at the right tail of the distribution include the equivalent areas at the left (hence the percentages here are in small type). Thus 80% of the two-tailed distribution lies between the two 10% markers, whereas in the one-tailed distribution 80% is included by the entire area under the curve to the left of the 20% marker.

So the less than 1 in 20 chance for variance ratios in the analysis of variance all lies in the tail of ONE half of the F distribution curve (Fig. 9.1). The whole of the other half of the curve is part of the 95% of the area that is **within** the critical limits. This contrasts with how we have considered 95% confidence limits up to now – as excluding 5%, yes, but a 5% made up of two 2.5% tails at the two opposite ends of the normal distribution (see Fig. 5.2).

The F-test in the analysis of variance is therefore a one-tailed test and tables of F are one-tailed tables!

The two-tailed F-test

If our F-test is two-tailed (as in testing for equality/inequality of two variances, see below) than the $P = 0.05$ values in the F tables are really at the higher probability of $P = 0.1$, and to truly test at $P = 0.05$ we have to use the tabulated values for $P = 0.025$. Not many people seem to know this! Also, you will need a more comprehensive table of F than provided here in Appendix A2.3.

BOX 9.1

The data in Box 8.6 of the individual tuber weight (g) of "Maris Piper" and "Jupiter" potatoes were used as an example of a t-test for two means with unequal variances.

The variances were 6.67 for 40 "Maris Piper" potatoes and 1.46 for 29 "Jupiter" potaoes.

Equality/inequality of variances is measured by the ratio of 6.67 (with 39 degrees of freedom) to 1.46 (with 28 d.f.).

This ratio of variances ($=F$, or variance ratio) is $6.67/1/46 = 4.57$.

Since there is no logical reason why "Jupiter" could not have had the greater variance, the variance for "Maris Piper" is only "on top" in the ratio in order to give an F greater than 1 to look up in the tables for 39 and 28 d.f. for the greater and smaller variances respectively, As (see main text) the F tables are in one-tailed form, we need to look at the figures for $P = 0.025$ to find the true F for $P = 0.05$ in a two-tailed test.

In tables giving F at the 0.025 level of probability, the tabulated value is 2.050. Our variance ratio is larger than this; we can conclude the variances depart significantly from equality.

You probably don't remember by now, but go back to page 70 where a test for equality/inequality of variances is needed to decide which type of t-test to use. We divide the larger variance by the smaller to get a variance ratio and check in F tables whether the ratio exceeds the tabulated F value at the two $n - 1$ coordinates for the two variances (see Box 9.1). If so, the variances are sufficiently unequal to justify using the form of t-test that can accommodate unequal variances. As for t values, the largest values of F that might arise purely by chance from a single normal distribution are available in tables. However, the F tables are "2 dimensional" in terms of degrees of freedom since there are different degrees of freedom associated with each of the two variances (Appendix A2.3).

But we have to remember that, when we use F tables to test for equality of variances, the test is a two-tailed one, because we are testing the difference in magnitude of two variances with no predetermined direction of difference. Yet the F tables in books are one-tailed (see earlier). So, in the test for equality of variances, the $P = 0.025$ figures in the tables are now the appropriate ones for the $P = 0.05$ level of probability, the $P = 0.05$ figures for true $P = 0.1$, etc.

How many tails has the *t*-test?

So far we have discussed one- and two-tailed tests in relation to the *F* distribution, which is normally tabulated for one-tailed tests. By contrast, tables of "*t*" are usually presented for a two-tailed probability distribution, and the examples of *t*-tests in Chapter 8 were all two-tailed tests. Thus, in comparing the statistics marks of men and women, the weight of potato tubers of two varieties, or the dry weight of plants in two soils, a result in either direction (e.g. men higher marks than women or *vice versa*) were theoretically possible statistically valid results.

However, this may not always be true with *t*-tests. For example, we may wish to test whether a caterpillar feeding on a leaf for only one hour is going to remove a statistically significant amount of leaf area. There is obviously no way it can make leaves bigger! So if the mean leaf area of leaves with a caterpillar is larger than the area of leaves without caterpillars, the most that can have happened (whatever the size of the difference resulting from the chances of sampling) is that the caterpillars have not eaten away any leaf area. The difference (however big) must be a bad estimate of a true zero difference. Only a leaf area difference "eaten" is less than "uneaten" can be real! "Uneaten" less than "eaten" has to be a nonsense!

Our data for the paired *t*-test (Box 8.8) could be from just such an experiment, where we have paired two leaves from each of ten plants – but now let us imagine that column A is the leaf without caterpillars and B is the leaf that is eaten to a greater or lesser extent depending on plant species.

Our calculated *t* for a **two-tailed** test for a comparison of Soils A and B was 2.262 for 9 d.f. at $P = 0.05$, exactly the same as the tabulated *t* (page 78). Had the data been for a **one-tailed** test (e.g. from the caterpillar feeding experiment), the same calculated *t* of 2.262 would have been tested for significance at $P = 0.05$ by looking up the tabulated *t* – still for 9 d.f. – but now in the $P = 0.1$ column of the table. Here *t* for 9 d.f. is lower at 1.833. Our calculated *t* is clearly greater than this, and we can reject the null hypothesis of no reduction in leaf area by the feeding of caterpillars with greater certainty than with the two-tailed test.

The greater likelihood of the one-tailed test showing a significant difference between means makes it tempting to invoke this test in preference to the two-tailed test. So can we argue that, with the data in Box 8.3 about the statistics marks, we were testing whether women gained **higher** marks than men? Well, I'm afraid that is just not on! The acid test for the validity of a one-tailed comparison is that, **before the data are collected**, we know we would have to regard a "significant" difference between the means in one direction as "no difference." Even if we were trying to show that women gained better marks than men, an answer that men gained more marks than women would have to be accepted, however reluctantly. Only if that answer were a logical impossibility are we justified in using a one-tailed test.

The final conclusion on number of tails

The take-home message is – if in any doubt at all, stick with a two-tailed test!

10

Analysis of variance – What is it? How does it work?

Chapter features

Introduction

Analysis of variance allows us to calculate the background variability (pooled s^2) in experiments with more than the two treatments which are the limit to our calculation of the pooled variance (s^2) in the t-test. However, the pooled s^2 so calculated can be doubled, divided by n and square rooted to the *standard error of difference between means* (or s.e.d.m. for short) just as in the t-test. So we finish up the analysis of variance again calculating the s.e.d.m. for evaluating differences between means by the t-test. Although, as we go further, you will find we can gain a lot of valuable insights before we reach the use of t, many statisticians I know argue that the whole purpose of the analysis of variance is to calculate just one number, the pooled variance (s^2), in order to calculate t for testing differences between means.

The essential difference between analysis with the t-test and analysis of variance is as follows: For the t-test, we calculate the background variability by ADDITION. We **add** (pool) the sum of squares of the biological material calculated separately for the two columns of numbers (page 65). In the analysis of variance we use SUBTRACTION. We first calculate the total variability of the data – including both background and intentional variation from treatments – and then **subtract** the variability from various sources until there is left a remainder (residual) which represents the background variability of the biological material. From this remainder we calculate a pooled s^2 which has contributions from all the treatments.

Sums of squares in the analysis of variance

The analysis of variance actually relies on partitioning the total variability into how much has come from the various sources of variability in the experiment, including the "remainder" referred to above, first as sums of squares.

You may remember from the t-test (page 64 and Box 8.5) that we could add squares of deviations from the mean together singly or in groups. Squared deviations (i.e. $\sum(x - \bar{x})^2$) can be added. That, on a calculator, we use the identity $\sum x^2 - (\sum x)^2/n$ (added squares – correction factor) doesn't alter the fact that we are adding individual $\sum(x - \bar{x})^2$ values, though it does rather obscure it!

If we can add squared deviations from the mean before dividing by degrees of freedom, then obviously they are equally fair game for subtraction, the principle distinguishing the analysis of variance (from now on we'll usually call it *Anova* for short) from the t-test.

However, we cannot similarly add or subtract ratios such as variance (i.e. sum of squares divided by degrees of freedom). That is why the Anova apportions variability to the various "sources" of variability as sums of squares **before** dividing by degrees of freedom.

Some "made-up" variation to analyze by Anova

Usually you have experimental data for Anova, and you have to trust the analysis to partition the variability between the various sources of variability correctly. However, there is nothing to stop us putting Anova to the ultimate test – we can invent numbers representing a combination of known variabilities and then invite Anova to sort it out. We can then check how well it has coped. In the process you will encounter **SqADS** for the first time, but will also be given the chance (if you wish to take it) to understand what is actually going on.

We'll begin with a table of random numbers (page 106) and pick 16 single digit numbers (i.e. between 0 and 9) for a pattern of four columns and four rows (Table A).

Table A

2	1	8	4	
5	4	0	8	
7	7	5	2	
5	7	9	5	Total = 79

The variability of these 16 numbers as the sum of squares$_{\text{of deviations from the mean}}$ (but using the added squares minus correction factor method of calculation is $2^2 + 1^2 + 8^2 + \cdots + 5^2 - 79^2/16 =$ **106.94**).

Now let's add some **nonrandom** (=*"systematic"* or *"imposed"*) variability and increase the numbers in the columns by a different fixed amount. So we'll increase all numbers in column 1 by 2 and the other columns by 4, 6, and 8 respectively (Table B):

Table B

4	5	14	12	
7	8	6	16	
9	11	11	10	
7	11	15	13	Total = 159

The sum of squares of Table B is $4^2 + 5^2 + 14^2 + \cdots + 13^2 - 159^2/16 =$ 192.94. So our column weightings have increased the sum of squares by $192.94 - 106.94 = $ **86.00.**

Now we'll add double these column weightings to the four rows, i.e. 4, 8, 12 and 16 respectively (Table C):

Table C

8	9	18	16	
15	16	14	24	
21	23	23	22	
23	27	31	29	Total = 319

The sum of squares of Table C, which is the sum of three sources of variability (random numbers, column weighting, and row weighting) is $8^2 + 9^2 + 18^2 + \cdots + 29^2 - 319^2/16 = 660.94$. Therefore the row weighting has added $660.94 - 192.94 = $ **468.00** to the total sum of squares.

The sum of squares table

We can now draw up a sum of squares table showing how the variation of the numbers in Table C has been assembled from the three sources of variation (Table D):

Table D

Source of variation	Sum of squares
Column weighting	86.00
Row weighting	468.00
Random	106.94
TOTAL	660.94

Using Anova to sort out the variation in Table C

Having created the confusion of variability represented by the numbers in Table C, we'll use Anova to sort it out again into its components. We'll divide the process into three phases, Phase 1, Phase 2, and an End Phase. These three phases apply to most analyses of variance, regardless of their complexity.

Phase 1

Note: The preliminaries (see start of Chapter 11) to an Analysis of Variance involve tabulating the data in a form suitable for analysis and setting out a skeleton Anova table, listing the sources of variation and allocating degrees of freedom. In this chapter only, however, our table of data is already suitable for analysis, and it makes more sense to tackle the Anova table later and a bit at a time.

Phase 1 has three steps:

1 Total both columns and rows, and check horizontally as well as vertically that both sets of end totals add up to the same *Grand Total*. This check is vital; it is so easy to make a mistake! Even a small error in decimal points could cause you no end of trouble later in the analysis.
2 Use the Grand Total to calculate the *correction factor* for the sum of squares of the numbers in the body of the table, and write it

down! This correction factor will be used repeatedly throughout the analysis.

3 Calculate the added squares for the numbers in the body of the table, and subtract the correction factor to obtain the "Total sum of squares" (of deviations from the Grand Mean, of course).

Doing Phase 1 for Table C gives us Table E. The original data are in italics, and 319 is the Grand Total:

Table E

	Columns				
	8	*9*	*18*	*16*	51
Rows	*15*	*16*	*14*	*24*	69
	21	*23*	*31*	*29*	89
	23	*27*	*31*	*29*	110
	67	75	86	91	**319**

Correction factor (CF) $= 319^2/16 = $ **6360.06**

Total sum of squares (a calculation we did earlier for Table C) $= 8^2 + 9^2 + 18^2 + \cdots + 29^2 - \text{CF} = $ **660.94**.

That is the end of Phase 1. Phase 2 is based on the totals at the end of the columns and rows, and we'll never need the italicized numbers in the body of the table again. Remember this last bit – it will help you work through Anovas smoothly in the future.

Phase 2

The version of Table E that we need for Phase 2 is therefore simplified to Table F by eliminating the original numbers in the body of the table, reinforcing the message that Phase 2 involves only the column and row end totals:

Table F

	Columns				
	–	–	–	–	51
Rows	–	–	–	–	69
	–	–	–	–	89
	–	–	–	–	110
	67	75	86	91	**319**

Correction factor (CF) $= 319^2/16 = $ **6360.06**

In Phase 2, we will use the totals in Table F to work out how much of our total sum of squares of 660.94 has come from the weightings we gave columns and rows. Once we know this, the contribution of the original random numbers will be the "unaccounted for" remainder – i.e. we obtain it by subtraction.

Remember, once we leave Phase 1 we have no further use for the original numbers in Table E, and use only **totals** of numbers in our sum of squares calculations. For example, the column variation is expressed by the four totals 67, 75, 86, and 91 (from Table F). We already have the correction factor (CF) of 6360.06. However, to use our four column end totals for the "added squares" bit, we have to remember they are **totals of four numbers** and divide the added squares by 4 before we subtract the correction factor (Box 10.1 explains why we need to do this). You will find that calculating the variability due to our row and column weightings – simulating the effect of treatments in biological experiments – is such a simple concept it is positively cheeky! Anyone, including you, could have invented Anova!

SqADS – an important acronym

It is time to introduce you to **SqADS**, which stands for **SQUARE, ADD, DIVIDE, and SUBTRACT.** This is a useful acronym for calculating sums of squares in Anova after the end of Phase 1. We **SQUARE** what are always **totals** of original numbers, **ADD** these squares together, **DIVIDE** by the number of numbers producing each of the totals we have squared, and finally we **SUBTRACT** always the same correction factor as we calculated for Phase 1. **SQUARE, ADD, DIVIDE**, and **SUBTRACT**. So the sum of squares from the column weightings in our example is calculated by **SqADS** from **the column totals** as:

$$\frac{67^2 + 75^2 + 86^2 + 91^2}{4} - CF \text{ (still } 6369.06)$$

giving the answer **87.69** as the sum of squares for column weighting.

Note: A reminder that the divisor of 4 is the number of numbers contributing to the column total (therefore actually the number or rows) and NOT the number of columns! Note also that the number of totals to be squared and added times the divisor is always the same throughout the analysis and equals the number of original numbers in the body of the table (16 in our example).

 These are two things you should try to remember for all time NOW!

BOX 10.1

Let's actually look at what is going on in calculating the sum of squares for the column weightings. What happens is that we are still going to have 16 numbers adding up the grand total of 319, but they are going to be different numbers from those in Table E. We are going to pretend that there is no random variability; also that there is no row variability (all the rows will therefore total the same). So the only information we are going to use is shown in Table G:

Table G

Rows	Columns				
	–	–	–	–	–
	–	–	–	–	–
	–	–	–	–	
	–	–	–	–	–
	67	75	86	91	**319**

Correction factor (CF) $= 319^2/16 = $ **6360.06**

If there is neither random nor row variability, then the four numbers in each column must surely all be the same, namely the column mean (i.e. the column total divided by 4 in our example) (Table H):

Table H

Rows	Columns				
	16.75	18.75	21.50	22.75	79.5
	16.75	18.75	21.50	22.75	79.5
	16.75	18.75	21.50	22.75	79.5
	16.75	18.75	21.50	22.75	79.5
	67	75	86	91	**319**

Correction factor (CF) $= 319^2/16 = $ **6360.06**

Note the absence of variability between the rows! The sum of squares for column weightings is the sum of squares of the 16 imaginary numbers (four each of the different column mean) in the body of Table H. It's as crude as that! i.e. $16.75^2 + 18.75^2 + 21.50^2 + \cdots + 22.75^2 - \text{CF} = $ **87.69**.

For the row weighting sum of squares, we would similarly insert the row means in the body of Table I and calculate the sum of squares of these new 16 numbers:

Table I

Rows	Columns				
	12.75	12.75	12.75	12.75	51
	17.25	17.25	17.25	17.25	69
	22.25	22.25	22.25	22.25	89
	27.50	27.50	27.50	27.50	110
	79.75	79.75	79.75	79.75	**319**

Correction factor (CF) $= 319^2/16 = $ **6360.06**

BOX 10.2

SqADS is just a short cut for having to square and add several sets of identical numbers (e.g. four identical numbers in each column in Table H). For column 1, our familiar "added squares – correction factor" calculation would begin with $16.75^2 + 16.75^2 + 16.75^2 + 16.75^2$, i.e. four times the square of 16.75.

To derive **SqADS**, we begin with the fact that 16.75 is also the total of column 1 divided by 4, i.e. 67/4.

So we can replace $16.75^2 \times 4$ by $(67/4)^2 \times 4$, and can re-write the latter as:

$$\frac{67}{4} \times \frac{67}{4} \times 4$$

The top 4 cancels out one of the bottom ones, leaving $67^2/4$. That's the **Sq D** bit! When we add the squares for all four columns, we are adding $67^2/4 + 75^2/4 + 86^2/4 + 91^2/4$, which is the same as adding all the squares first **(SqA)** and then **D** (dividing) that total by 4, leaving just the **S** (subtracting the correction factor) to complete **SqADS**.

So for column totals, **SqADS** is:

$$\frac{67^2 + 75^2 + 86^2 + 91^2}{4} - CF$$

Statistics is full of computational short cuts which obfuscate what you are really doing. Box 10.1 shows that the **SqADS** procedure is derived from (i) adding squares after replacing the original data being analyzed by several identical mean values and then (ii) subtracting the correction factor. This already is one step removed from what it is calculating – the sum of the squared deviations from the overall mean of the data.

Box 10.2 shows how we derive **SqADS** for column totals from the familiar formula for added squares minus CF for the 16 numbers in the body of Table H of Box 10.1.

To get the sum of squares for row weightings, we similarly **SqADS** the row totals in Table F (replacing the squaring and adding of several identical row means in Table I of Box 10.1). The calculation is:

$$\frac{51^2 + 69^2 + 89^2 + 110^2}{4} - CF \text{ (still } 6369.06)$$

giving the answer 485.69. We have therefore calculated that, of a total sum of squares for Table C of 660.94, the column and row weightings account for 87.69 and 485.69 respectively. The remainder of 87.56

(i.e. $660.94 - 87.69 - 485.69$) is our estimate of the random number variation which started it all off in Table A.

Back to the sum of squares table

We are now in a position to compare the variability we put in with what Anova has dissected out. You'll see in Table J that we do this at the "sum of squares" level – I hope you remember why this is so? It's because we can no longer add or subtract variability once we have divided sums of squares by their degrees of freedom (page 64 and Box 8.5).

Table J

Source of variation	Sum of squares (ACTUAL – from Table D)	Sum of squares (by Anova)
Column weighting	86.00	87.69
Row weighting	468.00	485.69
Random	106.94	by subtraction 87.56
TOTAL	660.94	660.94

How well does the analysis reflect the input?

I suppose that there are two reactions you may have to Table J. Hopefully, you are amazed how well the sums of squares from the Anova reflect how the numbers in Table C were set up. Alternatively, you may be scornful that the match is not exact.

Unfortunately, Anova has no way of knowing what the actual weightings were – it can only do its best! The random sum of squares is what is left over when Anova has made its best guess for column and row weightings based on the only information available to it, namely the column and row totals. This is the real world we face when using Anova on the results of our biological experiments: we are never (as here) in the privileged position of knowing in advance how the variation has been assembled! In Anova the "systematic variation" (column and row weightings) "capture" a little of the random variation since the systematic variation is calculated first (Box 10.3).

BOX 10.3

To understand how Anova can overestimate the systematically introduced column and row weightings at the expense of the random variation, just look at the random variation of the first two rows in Table A. By chance, the rows add up to different numbers, 15 and 17 in this example.

This random +2 of row 2 is part of the reason why, by the time we reach the final numbers in Table E, row 2 adds up to 69, 18 more than row 1 at 51.

Anova then guesses the row variation from the row totals alone (Table F) by assuming that the 18 extra in the row 2 total should be equally distributed at $18/5 = +4.5$ in each of the four columns.

However, if you go back to Table C, you'll note that the true extra weighting given to row 2 over row 1 was not +4.5, but only +4.0. Anova has no way of knowing this, and ascribes the extra +0.5 to the sum of squares for rows instead of the rightful sum of squares, that for random variation.

End Phase

Note: *Why not Phase 3? Answer: All Anovas have Phases 1 and 2 and an End Phase, but in some (see later) there is a Phase 3 to come!*

In actual experiments, the columns and rows of Table C would probably represent, respectively, our experimental treatments and four repeats of each (called *replicates*). In the End Phase, we compare the magnitude of such imposed variation with the random variation in the form of *mean squares* (actually variance) rather than sum of squares. We therefore divide the sums of squares by the appropriate degrees of freedom (see below).

Degrees of freedom in Anova

In Anova, just as the sums of squares for the different sources of variation add up to the total sum of squares, so also do the degrees of freedom.

Now remind yourself from page 16 that degrees of freedom represent the "opportunities for variation" given that our calculations are always based on a sample of a potentially much larger pool of numbers. Using the mean (or total) of our sample to represent the true mean for our sum of squares calculations constrains the freedom for variation in our observations. The

Grand Total of 319 (or Grand Mean of 19.94) is the only constraint on the freedom for variation among the 16 numbers of Table C. Given any 15 of our 16 numbers in addition to the Grand Total, we know what the 16th number must be! So the Total d.f. are 15, and the d.f. for the three sources of variation will add up to this number.

What are the constraints on the freedom for variation of the column weightings? The four column totals in the **SqADS** procedure reflect that only four different numbers (apart from the grand total) appear in Table G in Box 10.1. The first three totals give us 3 d.f. In fact, that's all the degrees of freedom there are, since the last total must bring the Grand Total to 319 and so is not a fourth opportunity for variation. Thus the d.f. for four columns turns out (conveniently to remember) to be 3, i.e. $n - 1$ for the number of columns. This is elaborated in Box 10.4.

A similar argument applies to rows, which therefore also have 3 d.f.

So 6 of our 15 total d.f. are accounted for by Columns and Rows, leaving the remaining 9 for the *residual* random variation. It is in fact no coincidence (see Chapter 12) that the d.f. for the residual random variation is 3×3, i.e. the product of the columns × rows d.f. Indeed, I recommend that you always calculate the residual d.f. as a product rather than by subtraction; it is a much better check that you have sorted out the degrees of freedom in an Anova correctly.

BOX 10.4

Here we repeat Table H from Box 10.1, showing how much each degree of freedom (opportunity for column weighting variation) – and framed in a black box – completes the 16 spaces in the table together with the Grand Total (which then accounts for the last four numbers as shown by the dotted arrow).

	Columns				
	16.75	18.75	21.50	22.75	79.5
	16.75	18.75	21.50	22.75	79.5
Rows	16.75	18.75	21.50	22.75	79.5
	16.75	18.75	21.50	22.75	79.5
	67	75	86	91	319

Thus the analysis table enlarges to:

Source of variation	Degrees of freedom	Sum of squares
Column weighting	3	87.69
Row weighting	3	485.69
Random	(3 × 3) 9	by subtraction 87.56
TOTAL	15	660.94

The completion of the End Phase

The final Anova table involves the addition of two further columns for "*mean square*" and "*variance ratio*." I'll first present the completed table below and then go on to explain these two additional columns (we do not need to calculate the values shown as "xx," since they are not needed).

Source of variation	Degrees of freedom	Sum of squares	Mean square	Variance ratio
Column weighting	3	87.69 →	29.23	3.00
Row weighting	3	485.69 →	161.90	16.64
Random	9	by subtraction 87.56	9.73	xx
TOTAL	15	660.94	xx	xx

Did you ever learn how to read grid references on a map? It is always hard to remember that the numbers along the bottom of the map come before the ones up the side! One way of reminding yourself is the little phrase "You go along the corridor before you go upstairs!" – It's got a sort of rhythm to it that makes it hard to forget (see the arrows on the above table).

So "You go along the corridor" is how we calculate the *mean square*. We work horizontally dividing each sum of squares by its degrees of freedom (e.g. 87.69/3 for columns). So *mean square* is the sum of squares per degree of freedom. It's as simple as that! Much earlier in this book (page 15) that's exactly what defined "variance." So mean squares are really just variances; hence "analysis of *variance*." The mean square for the residual sum of squares (the original random numbers in our example) – the *residual mean square* – is our s^2, equivalent to the pooled variance of the *t*-test, and from which we calculate our s.e.d.m. for looking at differences between means.

"You go upstairs" reminds us that the *variance ratio* involves working vertically in the table. The *variance ratio* is calculated, for all the mean squares relating to sources of systematic/imposed variation, by dividing them by the residual (random) mean square.

The variance ratio

The *variance ratio* (F) tells us whether, if at all, our experimental treatments ("column" and "row" weightings in this chapter) cause variation in the numbers greater than the chance residual variation. Put crudely – "Can we hear the effect of our treatments over the background noise?"

An F of less than 1 indicates that the effect of our treatments is even less than the variation that happens by chance! It is therefore negligible in biological terms, and only F values well in excess of 1 are worth checking for significance (a one-tailed test, see Chapter 9). We check the tabulated values of F (Table A2.3) to see whether our calculated variance ratio (F) is sufficiently large for us to reject the null hypothesis, i.e. it is **unlikely** that we could have obtained such a large value of F by chance sampling of the material without any imposed variation (e.g. treatments) applied by us (Box 10.5).

As usual, we use the $P = 0.05$ values in the table. We find the correct degrees of freedom for our imposed variation along the top of the table, and

---

BOX 10.5

Table A in this chapter was 16 random numbers laid out as a 4 × 4 grid. If analyzed by Anova, chance positioning of larger numbers results in variance ratios for Columns and Rows; i.e. the analysis does not apportion all the variation to chance. Admittedly, the variance ratios are very small.

Source of variation	Degrees of freedom	Sum of squares	Mean square	Variance ratio
Column weighting	3	1.69	0.56	0.06
Row weighting	3	17.68	5.90	0.61
Random	9	by subtraction 87.56	9.73	xx
TOTAL	15	106.94	xx	xx

run down the column of tabulated values until we cross with the degrees of freedom for the residual variation. For both "Columns" and "Rows" in our example, the tabulated value of $F_{P=0.05}$ for 3 (along) and 9 (down) d.f. is 3.86. Our calculated F for "Columns" was only 3.00, so we cannot reject the null hypothesis that effectively there were no detectable column weightings. Of course in this example we know we put a systematic column weighting into the data, but it has just been obscured by the magnitude of the random variation. By contrast, the F for rows is 16.64, well in excess of the tabulated 3.86. If we use more stringent probability criteria, we will find that F for $P = 0.001$ (decreasing the probability of erroneously rejecting the null hypothesis from 1 in 20 to 1 in 1000!) is only 13.8. We can reject the null hypothesis that our F of 16.64 could merely be a chance result with confidence at $P < 0.001$. And, of course, the row weightings I added were twice as large as the column weightings. Anova has detected this with ease!

The relationship between "t" and "F"

If you go back to Box 8.4 in Chapter 8 and look at the data for the statistics test again, you will see that we have two "treatments" (men and women) and a number of "replicates" of each. In terms of Anova we have two "columns," each with a different number of rows (12 and 14 respectively). The data could therefore be analyzed just as well by Anova. The unequal number of replicates (rows) is dealt with in **SqADS** (see page 121) by **D**ividing by different numbers the two **Sq**uared totals (for men and women) before we **A**dd (See Box 10.6 for the calculations involved).

*Note: In this example, the replicates are **not** paired (page 75). So the two replicates in each "row" in the Anova have not finished up in the same row for any particular reason – there is no systematic row variation. The data would therefore be analyzed by Anova as for a fully randomized design (page 117).*

Just as with the t-test (where $t_{24\text{ d.f.}}$ was only 0.53), the null hypothesis, that there is no difference in the mean mark between the genders, seems confirmed by the tiny variance ratio ($F_{1\text{ and }24\text{ d.f.}}$) of 0.30. The t value of 0.53 is close to the square root of the F value ($\sqrt{0.30} = 0.55$). Another way of putting this is that $t^2 = F$. The t-test is always between just two treatments, but you can confirm this relationship between t and F (Box 10.7) by comparing t values (two-tailed, see Chapter 9) with the first column (i.e. for 1 d.f.) of F tables (one-tailed).

BOX 10.6 See Box 8.4 for the "Statistics examination" data

Correction factor (CF) $= \text{Total}^2/n = (709 + 865)^2/(12 + 14) = 95{,}287.54$
Total sum of squares $= 75^2 + 72^2 + 68^2 + \cdots + 39^2 - \text{CF} = 3778.46$

$$\textbf{SqAD} = \frac{\textbf{Sq}\text{uared}}{\textbf{D} \text{ by } n} \text{ and } \textbf{A}\text{dded totals} = \frac{709^2}{12} + \frac{865^2}{14} = 95{,}334.73$$

"Treatment" (=Gender) sum of squares $= 95{,}334.73 - 95{,}287.44 = 47.29$

Anova table:

Source of variation	Degrees of freedom	Sum of squares	Mean square	Variance ratio
2 Genders	1	47.29	47.29	0.30
Residual	24	by subtraction 3,731.17	155.47	xx
TOTAL (26 numbers)	25	3,778.46	xx	xx

BOX 10.7

Degrees of freedom	Tabulated t (2-tailed)	t^2	Tabulated F (1-tailed)
$P = 0.05$ (1 in 20 chance)			
4	2.776	7.706	7.71
12	2.179	4.748	4.75
20	2.086	4.351	4.35
$P = 0.01$ (1 in 100 chance)			
4	4.604	21.197	21.20
12	3.055	9.333	9.33
20	2.845	8.094	8.10
$P = 0.001$ (1 in 1000 chance)			
4	8.610	74.132	74.14
12	4.318	18.645	18.64
20	3.850	14.823	14.82

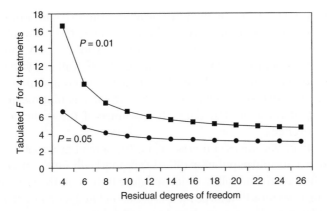

Fig. 10.1 F values for 4 treatments $(n-1=3)$, graphed from tables of F for $P=0.01$ and $P=0.05$, to illustrate the high values of F at low residual degrees of freedom and their decline with increasing residual d.f. till they change only little with 16 or more residual d.f.

Constraints on the analysis of variance

Adequate size of experiment

That we could not detect the column weighting in the size of the variance ratio is partly due to the few numbers, only 16, in our analysis, giving us only 9 residual degrees of freedom.

In laying out experiments, we need to give thought to how large the experiment needs to be. Unless we can get a reasonably good measure of the chance variation across the experiment, even a large Variance Ratio for a treatment effect may not exceed the tabulated F. Figure 10.1 shows how the F needed to reach significance at $P=0.01$ and $P=0.05$ declines with increasing residual degrees of freedom. The graph suggests that a good rule of thumb is to aim at 16 d.f. for the residual sum of squares. I shall often be using rather less than this in the examples in this book, but this is on purpose to make the calculations shorter!

Equality of variance between treatments

The residual mean square represents the background variance in the data, pooled across all "treatments" and replicates. We use it as the divisor for the mean squares of our imposed treatments, to measure how much larger the variation due to treatments is than the background variation. It is also the s^2 we use to evaluate differences between means. It is therefore important that it is a fair and equal measure of variability for **all** treatments.

If one treatment has atypically low variability (an extreme would be that all replicates show a value of zero) then – because the residual averages the variance across all treatments – we underestimate the variance with respect to the other treatments. This leads to a falsely inflated variance ratio (F) for treatments and a falsely small variance (s^2) in our significance tests of differences between means. It is therefore worth calculating, or even just "eyeballing" if there are few data, the variance of the replicates of each treatment to check.

Inequality of variances can arise when the spread of variability around the mean in a normal population differs very markedly between treatments, or when the data do not follow a normal distribution. In the latter case, it is usually fairly obvious that large means are associated with larger variances than small means. The remedy here is transformation (page 38) to normalize the data. If there is a different spread of variability between treatments, then a statistician should be consulted. If there is just one aberrant treatment, the simplest solution is to omit it from the analysis, but calculate the confidence limits (page 63) of its mean based on just its own variance. Then using the confidence limits based on the s^2 from the Anova of the remaining means, it is possible to see if the limits overlap.

Testing the homogeneity of variance

Chapter 9 described the two-tailed variance ratio test for equality/inequality of variances between the two treatments involved in a t-test. Once more than two treatments are involved, the solution is "*Bartlett's test for homogeneity between variances*," details of which will be found in more advanced textbooks on statistics. But be warned, it's a complex calculation! Fortunately there is a less accurate "rough and ready" method worked out by the distinguished statistician H.O. Hartley, where the value of the ratio of variances needed to show significant inequality increases as the number of treatments increases. The table (Appendix A2.4) is used with the extreme largest and smallest variances in the experiment.

The element of chance: randomization

Experiments are usually "laid out" in space as a number of replicate units (e.g. potted plants, field plots, aquaria, Petri dishes) of a number

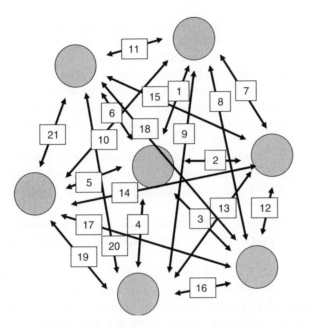

Fig. 10.2 The comparison of only seven numbers already involves 21 tests!

BOX 10.8

The number of tests for a given number of treatments is well represented by a "matrix" table. Say we have eight treatments A–H, the number of tests between treatment means that is possible is shown by the + symbol in the table below:

	A	B	C	D	E	F	G	H
A		+	+	+	+	+	+	+
B			+	+	+	+	+	+
C				+	+	+	+	+
D					+	+	+	+
E						+	+	+
F							+	+
G								+

Set out like this with 8 columns and $8 - 1 = 7$ rows, the number of tests between means (+) occupies exactly half of a 8×7 matrix. A general formula for the number of tests possible given any number of treatments is therefore:

$\frac{1}{2}$ (number of treatments $\times$ 1 less than the number of treatments) of treatments, the latter representing our imposed variation. Such layouts can follow a number of patterns, and the following chapters explore these patterns and how they may be analyzed by Anova.

However, the patterns all have one thing in common. This is that there should be no systematic pattern in the background variability contributing to the residual mean square of the Anova; systematic patterns are the prerogative of our imposed treatments or blocking (page 121) of the replicates (akin to pairing in the *t*-test)! We zealously preserve randomness of the background (residual) variation by letting the final arbiter of where "events" (i.e. the application of our different treatments) occur be pure chance. This process of allocating "events" to "locations" is called *randomization*.

If we have only a few treatments to randomize, then the easiest approach is to put the appropriate number of coded "raffle tickets" into a hat and make a draw!

If there are rather a lot of events we need to allocate, then probably the only truly random approach is to allocate numbers to "locations" and then select them randomly using "random numbers." Numbers which are very close to random selections can be obtained from computers by accessing the processor clock at intervals; it is not difficult to write an appropriate program for this and most statistical computer packages will already include a "random number generator."

Otherwise we seek refuge in published random number tables. These tend to appear as columns and rows of paired digits between 00 and 99. The numbers may be used in sequence as singletons (1–10, where 0 is taken as 10), pairs (01–100, where 00 = 100), threes (001–1000, where 000 = 1000), etc. The sequence followed may be down columns, along rows, or even diagonally. One may start anywhere in the table (not always at the beginning!) and the starting point should be chosen haphazardly, if not strictly by chance. One haphazard approach is to hit a pair of digits blind, and to use the first digit to identify the column and the second the row in that column where selections should start. Some folk are so fanatical, they will have some rule such as making the first digit the column co-ordinate on odd days of the month and the row co-ordinate on even days!

The process can be very tedious if one is trying to select a limited range of numbers from a much larger field (e.g. to randomize the numbers 1–17 from a possible 1–100), by leaving out numbers outside the range. It can take for ever to find the last few! The way round this (using our 17 out of 100 example) is to work out the highest multiplier of 17 which can be accommodated within 100; for 17 this would be $17 \times 5 = 85$. We can now use all pairs of digits up to 85 (therefore only discarding 86–00) by subtracting the nearest lower multiple of 17, and using the remainder as our random number. Thus the pairs 01–17 give us random numbers 1–17 and so do the ranges 18–34 (subtracting 17), 35–51 (subtracting 34), 52–68 (subtracting 51), and 69–85 (subtracting 68). As just one example, take the selection 62. The nearest lower multiple of 17 is 51, so our random number is $62 - 51 = 11$.

Comparison between treatment means in the analysis of variance

At the end of Anova, a significantly high Variance Ratio (F) for a source of variation (e.g. our experimental treatments) tells us that there is at least one significant difference to be found between the relevant means. It certainly does **NOT** tell us that all means differ from each other! For example, we may have measured the yield of four varieties of broad beans and found a significant F for "varieties." We know a significant difference lurks somewhere, but it could be that there is one high- or low-yielding variety, with the other three much of a muchness.

This is where we return to the t-test; It's a long way back on page 58. It will probably pay you to read this section of the book again, but for those given to laziness here is a quick recap. The t-test enables us to decide how likely a difference we have found between two means of n numbers could have arisen by chance sampling of the background variation. If that chance is less than 1 in 20 (the 5% chance, or Probability) $= 0.05$ where $P = 1$ is 100% certainty) we do not accept the null hypothesis that the difference we have found has arisen by chance and is just a poor estimate of zero. We can be more rigorous, and most statistical tables of t (and F) also give values for other levels of probability, e.g. for the 1 in 100 ($P = 0.01$) and the 1 in 1000 ($P = 0.001$) chances.

We regard a difference between means as statistically "significant" if our calculated value of the statistic "t" exceeds that in the tables for the relevant degrees of freedom at the chosen level of probability (usually $P = 0.05$ or less).

We calculate the value of the statistic "t" from the formula:

$$\frac{\text{Difference between the means}}{\text{s.e.d.m.}}$$

where s.e.d.m. is the *standard error of difference between means* (see page 45). Remember the mnemonic on page 52? – s.e.d.m. is the **S**quare root of the **E**stimated variance (the background variance of individual observations), after **D**oubling it and "**M**ean-ing" it (i.e. dividing by n).

In Anova, once we have subtracted the sums of squares for systematic variation (e.g. treatments) from the total sums of squares, the residual sum of squares measures the background variation of the individual observations. Dividing the residual sum of squares by its degrees of freedom gives us the *residual mean square* (page 99) and this is the "estimated variance" we can plug into our calculation of the s.e.d.m.

Let's leap ahead to page 127, where we have an Anova of an experiment where days to flowering of beans have been measured on four plots (replicates) per each of three fertilizer treatments:

Source of variation	Degrees of freedom	Sum of squares	Mean square	Variance ratio
3 Fertilizers	2	224.0	112.0	13.7
4 Replicates	3	441.7	147.2	18.0
Residual	6	by subtraction 49.3	8.2	
TOTAL (12 plots)	11	715.0		

The s.e.d.m. for fertilizer means (each of four plots) is thus:

$$\sqrt{\frac{2 \times 8.2}{4}} = \sqrt{4.1} = 2.02$$

The standard *t*-test (page 69) would then evaluate the difference between two fertilizer means as:

$$\text{is } \frac{\text{mean A and mean B difference}}{\text{s.e.d.m.} = 2.02} > t$$

(at $P = 0.05$ and 6 degrees of freedom)?

where we use 6 degrees of freedom because this is the number of d.f. on which the residual mean square we have used for calculating the s.e.d.m. is based.

The least significant difference

With three fertilizers, we would need to make three such *t*-tests for fertilizers A–B, A–C, and B–C, and with more treatments still this would become laborious. So the *t*-test in Anova usually takes the form of re-writing the above equation to calculate the smallest difference between means (of *n* numbers) that would be significant (at $P = 0.05$ and at the residual degrees of freedom). We therefore look up the tabulated *t* value needed for a difference to reach statistical significance, and use this in the equation above to calculate the size difference needed to reach that t value. This calculated difference is known as the *least significant difference*. We normally contract this to *LSD*, an acronym introduced long before the hallucinogenic drug bearing the same initials!

The equation to solve for the LSD is therefore:

$$\text{LSD} = t_{(P=0.05,\ \text{residual d.f.})} \times \text{s.e.d.m.}$$

or

$$\text{LSD} = t_{(P=0.05,\ \text{residual d.f.})} \times \sqrt{2 \times \text{residual mean square}/n}.$$

In our fertilizer experiment, the tabulated t for $P = 0.05$ at 6 d.f. is 2.447, so our LSD is 2.447 × the s.e.d.m. of 2.02 = **4.94**, and any difference between two fertilizer means in our experiment as large or larger than 4.94 is "statistically significant" at $P \leq 0.05$.

However, how we use the LSD test in experiments with several treatments is not that simple. We are making more than **one test between two variances**, so the probabilities (e.g. the 1 in 20 chance) no longer apply. With seven treatments, we are making 21 tests (see Fig. 10.2 and Box 10.8), so we are likely to find a "1 in 20 chance" significant difference where none really exists. Seven fertilizer means would equally have involved 21 tests as shown in Fig. 10.2, and the extreme largest and smallest means are likely to show an apparent significant difference, even if none exists and all seven means are really estimates of the same true average value. To illustrate the reality of this danger, we need to go back to page 46 and our several values for the mean weight of samples of 3 eggs.

If you go back to page 51, you will see that the s.e.d.m. of means of 3 eggs was 1.3. Since the variance on which this was based was calculated from 100 individual eggs, we look up the tabulated t for $P = 0.05$ at 99 d.f. and find it to be 1.98. So the Least Significant Difference for means of 3 eggs is 1.98 × 1.3 = **2.57**.

On page 46 you will find lots of means of 3 eggs drawn from the same normal distribution, i.e. they are all estimates of the same number – the true mean weight of all the eggs in the distribution. A group of seven means, in descending order of magnitude, were:

$$66.33, 65.33, 65.00, 65.00, 64.67, \text{ and } 63.67.$$

All 7 means were drawn by chance sampling of one normal distribution of weights involving just their natural variation with no imposed treatments, yet the largest and smallest means differ by 2.66, greater than our least significant difference between such means of 2.57. In analyzing an experiment where such means represented 7 treatment means we might well have concluded that the extreme means differ significantly because of our treatments, yet we have just seen that this "pseudo"-significant difference has been generated by chance sampling within one population of numbers.

A caveat about using the LSD

The whole question of using the LSD for significance tests in Anova needs a chapter to itself. The next chapters look at designs for Analysis of Variance and how to analyze them. With each design, we shall stop before the stage of LSD tests. However, we will not dodge the issue, but later (Chapter 16) return to the data, to show how we might use the LSD to identify the valid differences between means in each experiment.

EXECUTIVE SUMMARY 4
The principle of the analysis of variance

In the t-test, we combined the "within treatment" variability as a **pooled variance** (s^2). From this, we calculated the appropriate standard error ($\sqrt{2s^2/n}$) for judging the "statistical significance" of the difference between two "treatment" means.

We pooled variances by adding the two sums of ... squares of deviations from the mean (i.e. TOP half of variance equation) of the two columns of numbers **BEFORE** dividing by the pooled degrees of freedom.

In the analysis of variance, we still aim to calculate a pooled variance (s^2) to represent the inherent variation over which we have no control (basically sampling or "background" variation), but this time we do it by first calculating the total variability of the numbers in the experiment, and then **SUBTRACTING** rather than adding SUMS OF ... SQUARES of deviations (again before dividing by degrees of freedom).

This enables us to cope with more than two treatments, and also to handle more complicated experimental designs (Chapter 11).

The first stage of the analysis of variance is really **analysis of SUMS OF ... SQUARES of deviations** (= added squares – correction factor). We first calculate the **total variability** (as sum of squares) of the data. We then calculate (it is possible!) the variabilities (again as sums of squares) arising from the sources of variability designed into the experiment (i.e. usually the columns and rows of the data table). That part of the total variability we cannot account for by our experimental design remains as the "**residual sum of squares**" which, when divided by the appropriate degrees of freedom, becomes our variance (s^2).

Thus we visualize the number sampled from any one plot to have been influenced by three factors. Let's take just one number, the number in row A and column 1 of, say, 79. If the overall average per plot of the whole experiment is 50, this number contributes $(79 - 50)^2 = 29^2$ to the total sum of squares.

This deviation from the overall mean of 29^2 is:

partly an effect of column A – how much we can calculate;
partly because it is in row 1 – how much we can calculate;
for unknown reasons – this is the residual variation.

The procedure is as follows

1 Make a table of the results with columns and rows; then total rows and columns; then calculate the **grand total**, checking that both row and column totals add up to it.

2 Calculate the correction factor = grand total^2/number of plots.

3 Calculate **total variability** as the sum of squares of the plot results (added squares – correction factor). **YOU NOW NEVER AGAIN NEED THE ORIGINAL DATA, AND MOVE ON TO USE THE END TOTALS OF ROWS AND COLUMNS.**

4 Calculate **column variability** by **SqADS** using the column totals (see below). This is equivalent to calculating the sum of squares of a new table, where each plot result is theoretically replaced by the mean value for the column it is in, i.e. we apportion the grand total in such a way that the ONLY variability is that between columns (i.e. there is no variation within any one column and there is no experimental error).

5 Calculate **row variability** by **SqADS** of the row totals, this being equivalent to calculating the sum of squares of another new table, where each plot result is replaced by the mean value for the row it is in, i.e. we apportion the grand total in such a way that the ONLY variability is that between replicates.

6 We subtract (**4**) + (**5**) from (**3**) to find out how much variability is "left over" = **residual sum of squares** (**6**).

7 We convert sums of squares to variance (= mean square) by dividing (**4**), (**5**), and (**6**) by the appropriate degrees of freedom:
(a) total degrees of freedom are (number of plots – (**1**));
(b) treatment degrees of freedom are (number of treatment means used in table (**4**) – (**1**));
(c) replicate degrees of freedom are similarly (number of replicate means used in table (**5**) – (**1**));
(d) residual degrees of freedom are (a) – (b + c) OR also (b) × (c).

8 F (variance ratio) is how much bigger the effect of treatments and replicates is than residual variation (i.e. the variance ratio for treatments and replicates is their mean squares divided by the residual mean square).

SqADS in the analysis of variance

After calculating the TOTAL sum of squares *of deviations*, we replace all the figures by the means for treatments or replicates based on totals. We can replace this procedure more quickly by **SqADS**.

1 **Sq**uare each figure and **A**dd the squares together.

2 **D**ivide by the number of figures contributing to each total.

3 **S**ubtract the correction factor.

Thus for three totals – 44, 36, 23 – each of four figures:

1 $44^2 + 36^2 + 23^2$

2 $\dfrac{44^2 + 36^2 + 23^2}{4}$

Note there are three totals divided by $4 \cdots 3 \times 4 = 12$ (the number of original data)

3 $\dfrac{44^2 + 36^2 + 23^2}{4} - \dfrac{103^2}{12}$

Note 3×4 is also the divisor of the correction factor.

Sums of squares of deviations – how far we have come since Chapter 3!

The concept of **VARIANCE** is based on averaging (but by $n - 1$ and not simply n) the sum of all the squared differences between numbers in a set of data and their mean value. This sum is called the SUM OF SQUARES OF DEVIATIONS, and very commonly just SUM OF SQUARES.

For our example, we'll use the five numbers 5.4, 7.3, 6.2, 5.3, and 6.8, which have a total of 31 and a mean of 6.2.

The five numbers have respective deviations from the mean of -0.8, 1.1, 0, -0.9, and 0.6.

However we calculate this, the answer we are looking for is the sum of the squared deviations from the mean, i.e. $0.8^2 + 1.1^2 + 0^2 + 0.9^2 + 0.6^2 = 0.64 + 1.21 + 0 + 0.81 + 0.36$. These squared totals add up to **3.02**, and this is the SUM OF SQUARES OF DEVIATIONS.

It is easy to forget this basic concept, because we use different ways of calculating the same result.

Our **first** change was to FORGET about deviations, and use the actual numbers instead ("added squares")!

This is $5.4^2 + 7.3^2 + 6.2^2 + 5.3^2 + 6.8^2 = 195.22$. A large figure, which now needs the subtraction of a CORRECTION FACTOR for having used the numbers instead of their differences from their mean. The correction factor is the mean of the squared total of our numbers, i.e. 31^2 divided by $5 = 961/5 = 192.20$. Our corrected sum of squares of deviations is now correct at 195.22 minus 192.20 = **3.02** again.

Our **second** change was the further step of simplifying the calculation of the ADDED squares in the situation when a group of numbers was identical, e.g. 5.4, 5.4, 5.4, and 5.4.

Instead of calculating $5.4^2 + 5.4^2 + 5.4^2 + 5.4^2$ as $29.16 + 29.16 + 29.16 + 29.16 = \mathbf{116.64}$, we begin with the total of $(4 \times 5.4) = 21.6$.

We can get our ADDED SQUARES by squaring this total and dividing by 4, i.e. we **SQ**UARE (totals), **A**DD (them together), and **D**IVIDE (by the number of identical numbers forming the total). 21.6^2 divided by 4 is again **116.64. This is the ADDED squares ONLY, i.e. it has not been CORRECTED by subtracting a correction factor. This subtraction is the final S of SqADS.**

11

Experimental designs for analysis of variance

Chapter features

Introduction

When I first came to Reading University in 1961, Professor Heath was growing seven species of alpine plants in hexagonal plots on a hemispherical mound (like a half football with hexagonal panels) so that every plant species was competing with each of the others on one of the sides of the plot. The experiment was a novel design for analysis of variance, though it proved impractical and was never published. This is a good example of what has recently been pointed out to me: that it was the development of the technique of analysis of variance by R.A. Fisher in the 1920s which stimulated researchers to develop a variety of experimental designs exploiting the flexibility and elegance of the statistical technique. This chapter describes some of the most common of these designs. In those designs dealt with in greater detail, the format will be as follows:

- Feature of the design.
- A sample layout showing randomization of treatments.
- An example of data from such a layout; followed by the analysis in "Phases":
  - *Prelims* (see Chapter 10, page 91) – (1) Unscrambling the randomization (i.e. arranging the table of data for the analysis). (2) Establishing the "lead line." (3) Setting out the skeleton analysis table with allocation of degrees of freedom.

○ *Phase 1* (see Chapter 10, page 91) – (1) Total both columns and rows, and check addition to the Grand Total both ways. (2) Calculate the Correction Factor (Grand Total^2/n). (3) Calculate the added squares for the body of the table and subtract the Correction Factor (CF) to obtain the Total Sum of Squares. Remember, the data in the body of the table will not be used again.

○ *Phase 2* (see Chapter 10, page 92) – Use only end totals. To obtain the sum of squares for the systematic imposed sources of variation which end totals represent, **SqADS** the appropriate end totals. Remember **Square**^the end totals^, **Add**^these squares together^, then **Divide**^by the number of data contributing to each end total^, and finally **Subtract**^the Correction Factor^.

○ *End Phase* (see Chapter 10, page 97) – (1) Work horizontally in the analysis table, dividing the sum of squares for each source of variation by the relevant degrees of freedom to obtain the mean squares. (2) Work vertically, dividing each mean square for systematic variation by the residual mean square to obtain the variance ratios (F). (3) Look up tabulated values for each variance ratio greater than 1, to determine the probability (P value) of an F of such magnitude arising by chance.

Fully randomized

This simple design has many advantages, but it is often ignored even where it would be appropriate and advantageous to use it. Imagine we wish to compare the time to flowering of broad beans treated with three different fertilizers (A, B, and C) and that we plan to sow four replicate plots for each fertilizer treatment. We simply let chance decide (perhaps by writing 4 As, 4 Bs, and 4 Cs on pieces of paper and drawing them out of a hat) which plots receive which fertilizer. With a rectangular plot and with the hat yielding in order B, A, C, B, A, B, . . . A, the plan might be as in Fig. 11.1.

An advantage of this kind of design is that all but the 2 degrees of freedom for the three fertilizers are retained for the residual sum of squares. The layout can be any shape and need not even contain an equal number of plots for each fertilizer. Things do go wrong in experiments, and you might not be able to record data from the bottom right A plot if, for example, a passing elephant had sat on it!

The main problem with a fully randomized layout is that the area may not be sufficiently uniform. In the above example, chance has caused three C plots to be rather strongly clustered top right. If, for example, the soil is better in that part of the field, then the earlier flowering there would

B	A	C	B
A	B	C	C
C	B	A	A

Fig. 11.1   *Fully randomized design*: An example of how four repeats of three treatments might be randomly allocated to the 12 plots.

B 43	A 35	C 29	B 38
A 38	B 40	C 31	C 30
C 38	B 39	A 42	A 40

Fig. 11.2   *Fully randomized design*: The layout of Fig. 11.1 with data added (days to flowering of half the broad bean plants per plot).

contribute a spurious superiority to fertilizer C. If our plots are not fixed (e.g. potted plants in a greenhouse or Petri dishes in the lab.), we can get round this problem by re-randomizing and repositioning the units at perhaps weekly intervals.

Data for analysis of a fully randomized experiment

The data in Fig. 11.2 are days to flowering of 50% of broad bean plants per plot.

Prelims

Unscramble the randomization so that each plot per fertilizer A, B, and C is in the same column (or row if you prefer). From now on, we shall be adding subscripts to the end totals for systematic sources of variation, to remind us how many data contribute to each total. This becomes increasingly helpful with more complicated designs, and identifies the divisor (D) in SqADS.

A	B	C	
35	43	29	107
38	38	31	107
42	40	30	112
40	39	33	112
155_4	160_4	123_4	GT = **438**$_{12}$

Now a second addition is the "lead line." Again this becomes more useful later, but it's a good habit to get into now. The "lead line" identifies the components of the experiment and here it is:

$$3 \text{ Fertilizers} \times 4 \text{ replicates} = 12 \text{ plots}$$

We use the lead line to produce the skeleton analysis table with degrees of freedom $(n - 1)$. Thus 12 data (the plots) have 11 d.f. of which three fertilizers account for 2 d.f. As the four replicates are located at random, all remaining d.f. $(11-2)$ are attributed to residual variation:

Source of variation	Degrees of freedom	Sum of squares	Mean square	Variance ratio	P
3 Fertilizers	2				
Residual	(11 – 2 =)9				
TOTAL (12 plots)	11				

Phase 1

The Correction Factor (Grand Total^2/n) is $438^2/12 = 15,987$
Added squares in the body of the table are $43^2 + 35^2 + 29^2 + \cdots + 40^2 = 16,238$
Therefore Total Sum of Squares $= 16,238 - CF = 251$

Phase 2

There is only one systematic ("treatment") source of variation, and therefore only one set of end totals to SqADS to calculate the sum of squares. Hence only the column totals in the data table are in large type, and have subscripts to indicate that the divisor for the squared and added column end totals is 4. Remember again, the divisor is the number of data in the table contributing to each end total. So the calculation is:

$$\textbf{SqA} 155^2 + 160^2 + 123^2 = 64,754 \quad \textbf{D}^{by4} 64754/4 = 16,188.5$$

$$\textbf{S}^{CF} 16,188.5 - 15,987 = 201.5$$

You may wish to remind yourself at this point of what we are actually doing in this calculation (Box 11.1).

BOX 11.1

The sum of squares$_{\text{of deviations from the mean}}$ (remember page 15 and the notation $\sum(x - \bar{x})^2$) for fertilizers can be obtained directly from the table below. Here each datum in the body of the table has been replaced by the mean days to flowering for plants given that fertilizer:

A	B	C	
38.75	40.0	30.75	
38.75	40.0	30.75	
38.75	40.0	30.75	
38.75	40.0	30.75	
			GT = **438**$_{12}$
155$_4$	160$_4$	123$_4$	Mean = 36.5

The sum of squares as squared deviations from the mean is $(36.5 - 38.75)^2 + (36.5 - 40.0)^2 + (36.5 - 30.75)^2 + \cdots + (36.5 - 30.75)^2$, and needs no correction factor!

We normally replace this rather laborious calculation with the "added squares – Correction Factor" identity ($\sum x^2 - (\sum x)^2 / n$ in notation) and therefore calculate:

$$38.75^2 + 40.0^2 + 30.75^2 + \cdots + 30.75^2 - 438^2/12$$

SqADS is basically the same calculation, but replaces each squared mean value with the squared totals added and divided by 4 (the number of data contributing to each total), finally subtracting the same correction factor, i.e.

$$\frac{155^2 + 160^2 + 123^2}{4} - \frac{438^2}{12}$$

The same answer (**201.5**) will be obtained by each of the three calculation procedures. Why not check this out?

The sums of squares are therefore partitioned as:

Lead line: 3 Fertilizers × 3 plots of each = 12 plots
Correction factor: 15,987

Source of variation	Degrees of freedom	Sum of squares	Mean square	Variance ratio	P
3 Fertilizers	2	201.5			
Residual	(11 − 2 =)9	by substraction 49.5			
TOTAL (12 plots)	11	251.0			

End Phase

Horizontally in the table, we divide the sums of squares for Fertilizers and the Residual by their degrees of freedom and then, vertically, we obtain the variance ratio (F) for Fertilizers by dividing the mean square for Fertilizers by the residual mean square:

Lead line: 3 Fertilizers × 3 plots of each = 12 plots
Correction factor: 15,987

Source of variation	Degrees of freedom	Sum of squares	Mean square	Variance ratio	P
3 Fertilizers	2	⟶ 201.5 ⟶	100.75	18.32	= 0.001
Residual	(11−2 =) 9	⟶ by substraction 49.5 ⟶	5.50		
TOTAL (12 plots)	11	251.0			

The tabulated value (Appendix 2, Table A2.3) for F for 2 (for Fertilizers – along the top) and 9 (for Residual – down the side) degrees of freedom at $P = 0.05$ is 4.26 and even at $P = 0.001$ it is only 16.41, still lower than our calculated F for Fertilizers. P is therefore <0.001, i.e. an F as large as 18.32 would occur well less than once in 1000 times if fertilizers had had no effect and our plot data were all merely samples of the same background variation.

We can therefore be pretty sure that the time to flowering of the beans is different with different fertilizers, but of course it could be that just one has made a difference compared with the other two (with no difference between them). The further tests needed to interpret the experiment fully are detailed in Chapter 16, page 226.

Had the bottom right-hand corner been sat on by my hypothetical elephant, the data would still be easily analyzable. Total degrees of freedom would reduce to 10 (only 11 plots remain). There are still three Fertilizers, so d.f. for that source of variation remain at 2, and the d.f. for the Residual would reduce to 8. The data of the remaining 11 plots (now totaling only 398) would be used to calculate the total sum of squares. The Fertilizer sum of squares would merely use a different divisor for Fertilizer A (with only three replicate plots remaining):

$$\frac{135^2}{3} + \frac{160^2 + 123^2}{4} - \frac{398^2}{11} = 190.22$$

Note that this sum of squares of 190.22 is very similar to that of 201.5 for Fertilizers in the complete experiment.

Randomized blocks

As hinted under the fully randomized design above, soil, temperature and light conditions often lead to gradients in the greenhouse or field which may themselves affect the experimental variable (e.g. yield of a crop) that we are recording.

In our "made-up" analysis of variance in Chapter 10, we not only measured the variation due to column weighting (equivalent to fertilizer treatment in the fully-randomized experiment above), but we were also able to separate from the residual variation any systematic row weighting.

Therefore, if we know the direction of any problem gradient in our experimental area, we can arrange our experiment in such a way that the gradient becomes part of the systematic sources of variation designed into the experiment. Any systematic variation can be separated from the residual variation by Anova. Making the gradient a systematic source of variation is done by grouping the replicates of each treatment into blocks in different positions along the gradient (Fig. 11.3).

So back to the experiment above comparing the date of flowering of broad beans treated with three different fertilizers (A, B, and C) with four replicate plots for each fertilizer treatment. To change this experiment from a fully-randomized to a randomized block design we put only one paper with A, B, or C into the hat, but now draw them by chance four times, returning them to the hat after each draw. Each draw provides a chance allocation of A, B, and C within one replicate *block* of the experiment. This

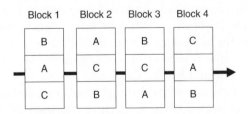

Fig. 11.3 *Randomized block design*: A possible randomization of three treatments to four blocks at right angles to a known gradient.

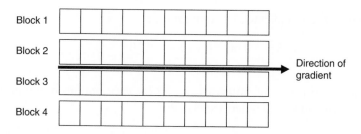

Fig. 11.4 *Randomized block design*: How not to do it! Four blocks of ten treatments aligned to make best use of a rectangular piece of land, but parallel with a gradient as a result.

gives a layout as shown in Fig. 11.3. Note that every treatment occurs once in each block. Thus *blocks* become replicate "experiments" of the same three treatments. The blocks are separated in Fig. 11.3 for clarity; there is no reason why blocks should not be contiguous if preferred.

The blocks can be any size or shape. The important thing is that they are laid out at right angles to the gradient that we suspect may influence the results. However, I have seen the same piece of land used in consecutive years with a layout like Fig. 11.3 in the first year then reversed to Fig. 11.4 in the second year accommodate more treatments.

This shows total ignorance of why blocking is used! If we are going to ignore real gradients, or don't know in which direction they go, there is no point in using this design. It is probably used so often without thinking because randomized blocks turn up in textbooks and research papers so frequently.

One of the real advantages of randomized blocks is that we can use the concept of "blocks" to separate out sums of squares for nonbackground variation other than field gradients. For example, with potted plants in greenhouses, it may pay us to "block for uniformity," putting all the largest plants into one block and so on to a block with all the smallest plants. Thus any size effect which might influence the results will be taken out in the block sum of squares. If space or time are limiting, we can use the block design to build up the required number of replicates by using several areas or completing several blocks at different times – again the space or time effect will go into the block sum of squares.

Now that each plot is subject to more than one systematic source of variation (treatment and block), the failure of one plot to yield a result cannot be ignored as it can with the randomized block design. If the number of blocks is large enough, a whole block can be "written off" even if only one plot fails, but it is usually more efficient to use a "missing plot" calculation. This calculation is outside the scope of this elementary book, but will be found in most larger texts; the principle is that a "best guess" is made of the

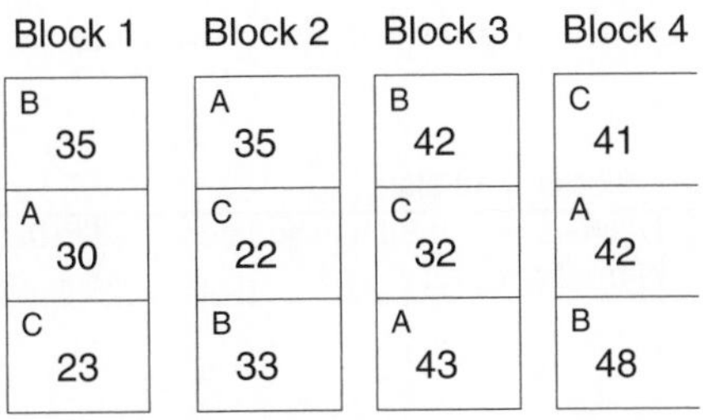

Block 1   Block 2   Block 3   Block 4

B 35	A 35	B 42	C 41
A 30	C 22	C 32	A 42
C 23	B 33	A 43	B 48

Fig. 11.5 *Randomized block design*: The layout of Fig. 11.3 with data added (days to flowering of half the broad bean plants per plot).

likely data that would have been recorded from one or two missing plots, with the sacrifice of some residual degrees of freedom in the Anova.

Data for analysis of a randomized block experiment

The data (Fig. 11.5) are again days to flowering of 50% of broad bean plants per plot.

Prelims

Unscramble the randomization so that each plot per fertilizer is in the same column (or row if you prefer). Note the subscripts added to the end totals for what are TWO systematic sources of variation in this design, subscripts are there to remind us how many data contribute to each total.

	A	B	C	
Block 1	30	35	23	88_3
Block 2	35	33	22	90_3
Block 3	43	42	32	117_3
Block 4	42	48	41	131_3
	150_4	158_4	118_4	GT = 426_{12}

Note: *The row (block) totals increase down the table, suggesting there is a real gradient to make the use of this design worthwhile.*

The "lead line" now uses the words "Blocks" rather than replicates to remind us that the replicates are now purposefully in "block" form, and is:

3 Fertilizers × 4 Blocks = 12 plots

We use the lead line to produce the skeleton analysis table with degrees of freedom $(n - 1)$:

Lead line: 3 Fertilizers $\times$ 4 Blocks $= 12$ plots

Source of variation	Degrees of freedom	Sum of squares	Mean square	Variance ratio	P
3 Fertilizers	2				
4 Blocks	3				
Residual	$(2 \times 3 =)6$				
TOTAL (12 plots)	11				

Compared with the fully randomized design for 12 plots analyzed earlier, you will see that 3 degrees of freedom have been lost from the residual and allocated to "Blocks." This reduces the divisor for the residual mean square, but this disadvantage of the randomized block design is more than compensated if the variance between blocks is considerably greater than the residual variation.

A second point on the table is how I have calculated the degrees of freedom for the residual variation. Although 6 d.f. could have been found by subtraction $(11 - 2 - 3 = 6)$, I have preferred (page 98) to multiply the d.f. for Fertilizers and Blocks. This is another way of arriving at the correct d.f. for the residual sum of squares, and has the advantage of being a much better check that we have allocated the other d.f. correctly.

It is no coincidence that multiplying the d.f. for Fertilizers and Blocks gives the correct residual d.f., because the residual variation represents what we call the *interaction* of fertilizers and blocks (which would often be written as the "Fertilizer $\times$ Block" interaction). You will hear a lot more about *interaction* in Chapter 12, where we become involved with interactions between different components of experimental treatments. Here, the interaction between Fertilizers and Blocks takes the form that sampling is unlikely to give data where the three different Fertilizers will each give the absolutely identical time to flowering of the beans in all four Blocks. This is described more fully in Box 11.2, but the point can also be made from the table of data (reproduced again below):

	A	B	C
Block 1	30	35	23
Block 2	35	33	22
Block 3	43	42	32
Block 4	42	48	41

You will see, for example, that in Block 2 Fertilizer C has advanced the date of flowering of the plants by 40% compared with Fertilizer A (shaded

BOX 11.2

If there were no interaction between Fertilizers and Blocks in our randomized block experiment. Then all three Fertilizer totals would have been partitioned between blocks 1–4 in the ratio 88:90:117:131 (i.e. according to the block totals). Thus the datum for Fertilizer A in Block 1 would not have been 30, but $150 \times 88/426 = 31$ (see bottom of this box*). The table where fertilizers are having a fully consistent effect in all blocks is then completed as:

	A	B	C	
Block 1	31.0	32.6	24.4	88_3
Block 2	31.7	33.4	24.9	90_3
Block 3	41.2	43.4	32.4	117_3
Block 4	46.1	48.6	36.3	131_3
	150_4	158_4	118_4	GT = 426_{12}

The added squares of these data come to 15,795.2. The total is still 426 giving us the same correction factor as on page 126 of 15,123.0 and therefore a total sum of squares of $15,795.2 - 15,123.0 = 672.2$.

Since the end totals have not changed, the Fertilizer and Block sums of squares are as for the original data (page 126) , 224.0 and 441.7 respectively. The sum of these, 665.7, leaves almost nothing (only 6.5) for the Residual sum of squares; it is the inconsistency in the original data in how the Fertilizers have performed relative to each other in the different Blocks (the "Fertilizer × Block interaction") which produced over 40 for the residual sum of squares in the original analysis. We should in fact have a residual of zero in the interaction-free analysis; the 6.5 comes from an accumulation of rounding up decimal places. For example, I have put the datum for Fertilizer A in Block 1 as 31.0 above, when the true answer is only 30.985915492958 . . .!

*Expected values with no inconsistency (interaction) are obtained *from column total × row total ÷ grand total*. If you need convincing that this is sensible, imagine that all 12 numbers in the table above are identical at the overall mean of 426/12 = 35.5. Then all column totals would be 35.5 × 4 = 142, and all row totals 35.5 × 3 = 106.5. Then every expected value would be 142 × 106.5 ÷ 426 = 35.5 ! . . . Yes!

numbers), yet in Block 4 (boxed numbers) the difference is negligible, only 1 day! So in a randomized block experiment, it is such inconsistencies which form our residual variation.

Phase 1

The Correction Factor (Grand Total$^2/n$) is $426^2/12 = 15,123$

Added squares in the body of the table are $30^2 + 35^2 + 23^2 + \cdots + 41^2 = 15,838$

Therefore Total Sum of Squares $= 15,838.00 - \text{CF} = 715$

BOX 11.3

Phase 2 sum of squares calculations using the column and row totals together with the correction factor already calculated in Phase 1.

$$\text{Fertilizer (columns) sum of squares} = \frac{150^2 + 158^2 + 118^2}{4} - 15{,}123 = 224$$

$$\text{Blocks (rows) sum of squares} = \frac{88^2 + 90^2 + 117^2 + 131^2}{3} - 15{,}123 = 441.7$$

Note that in both cases the number of totals squared and added $\times$ the divisor $= 12$, the number of plots in the experiment.

Phase 2

I hope you remember that, having calculated the total variability of the 12 numbers in the body of the table, we have now *finished with those numbers* and Phase 2 uses only *the column and row end totals* in its calculations. Hopefully, by now, you are also familiar with the SqADS procedure for calculating the sums of square for Fertilizers (columns) and Blocks (rows). Therefore, only a reminder follows below, but – for those who would still find it helpful – the full numerical calculations are given in Box 11.3:

$$\text{Sum of squares for Fertilizers} = \textbf{SqA}^{\,3\text{ Fertilizer totals}}\,\textbf{D}^{\text{by }4}\,\textbf{S}^{\text{Correction Factor}} = 224$$

$$\text{Sum of squares for Blocks} = \textbf{SqA}^{\,4\text{ Block totals}}\,\textbf{D}^{\text{by }3}\,\textbf{S}^{\text{Correction Factor}} = 441.7$$

Reminder: *The number of totals squared and added times the divisor $=$ the total number of "plots" $(=12)$.*

Thus the sums of squares are partitioned as:

Lead line: 3 Fertilizers $\times$ 4 Blocks $= 12$ plots
Correction Factor $= 15{,}123$

Source of variation	Degrees of freedom	Sum of squares	Mean square	Variance ratio	P
3 Fertilizers	2	224.0			
4 Blocks	3	441.7			
Residual	$(2 \times 3 =)6$	by subtraction 49.3			
TOTAL (12 plots)	11	715.0			

End Phase

Finally, we divide the sums of squares for Fertilizers, Blocks, and the Residual by their degrees of freedom and then obtain the variance ratios (F) for Fertilizers and Blocks by dividing their mean square by the residual mean square. The last step is to look up the P (probability) value for the two variance ratios in tables of F (with d.f. for Fertilizers [2] or Blocks [3] across the top and the 6 Residual d.f. down the side):

Lead line: 3 Fertilizers × 4 Blocks = 12 plots
Correction Factor = 15,123

Source of variation	Degrees of freedom	Sum of squares	Mean square	Variance ratio	P
3 Fertilizers	2	⟶ 224.0	⟶ 112.1	13.7	= 0.01
4 Blocks	3	⟶ 441.7	⟶ 147.2	18.0	= 0.01
Residual	(2 × 3=) 6	⟶ by subtraction 49.3	⟶ 8.2		
TOTAL (12 plots)	11	715.0			

Both calculated variance ratios are so large that they would occur well less than once in 100 times ($P < 0.01$) if we were merely sampling the background variation.

The further tests on the means needed to interpret the experiment fully are detailed later in Chapter 16, page 227.

Incomplete blocks

The number of treatments we wish to include in an experiment may be very large, particularly if it is a complex *factorial* experiment (see Chapter 12), where a number of treatments are each applied at several different levels (e.g. three fertilizers each applied at two rates on three varieties each grown in two different composts – giving $3 \times 2 \times 3 \times 2 = 36$ treatments!). Not only may there be insufficient space to accommodate several blocks of so many treatments, but the recording of the data may be too large a task. Moreover, the larger the area occupied by an experiment, the greater the variation in environmental conditions at the extremes of each block and therefore the larger the residual sum of squares in comparison to treatment effects.

Thought needs to be given to avoiding too many treatments, and it may be possible to define the aims of the experiment more clearly and eliminate some of the proposed treatments. Using a *split-plot* design (see later and Chapter 15) may also reduce the area needed for the experiment.

If there really is no alternative, then refuge can be taken in Incomplete Randomized Block or Lattice designs. Complex analyses are the

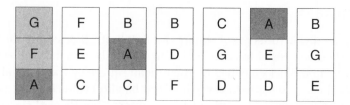

Fig. 11.6   *Incomplete randomized block design*: A possible layout of seven treatments in blocks of only three plots. Treatment A occurs three times (dark shading), as do all other treatments. All pairs in the same block (e.g. G and F in the first block) also occur the same number of times (just once in this layout).

consequence of choosing such designs, and such analyses are outside the scope of this book. Further advice should be sought; they are introduced here to make you aware that solutions to accommodating a large number of treatments are available.

Figure 11.6 shows an Incomplete Randomized Block layout, each of only three plots but accommodating seven treatments. The 21 plots involved could of course have been three blocks of the seven treatments, but using seven blocks gives more efficient control of residual variation. The criteria required for analyzing such a design is that each treatment appears the same number of times (three in our example – Fig. 11.6 , e.g. A) and that so does each pair found together in the same block (here just once, e.g. G and F). The treatments assigned to each block by these criteria are then randomized within the blocks.

Lattice designs have different criteria and are applicable to large factorial (see later) experiments involving many interactions between treatments. One particular difference from incomplete randomized blocks is that a pair of treatments appearing together in the same block shall **not** coincide in other blocks. A typical lattice design for nine treatments, with four replicates (there should be an even number), is shown in Fig. 11.7. Note that each replicate itself has three treatments in two dimensions, as columns and rows (the blocks), and these dimensions feature in the complex analysis. The treatments are allocated to the first replicate at random. In the second replicate, the rows become the columns, and this fulfills the criteria. Thus two replicates (replicates 1 and 2 below) accommodate six blocks of the nine treatments in 18 plots compared with the 54 plots which would be needed in a complete randomized block design. Further replicates (e.g. replicates 3 and 4 below) can be created by imagining the lattice ever repeating in both dimensions, and using diagonals starting at different points. You will see the diagonal G, H, and E of replicate 1 appearing as the first block of replicate 3; B, I, and D is the next diagonal transferred from replicate 1 to

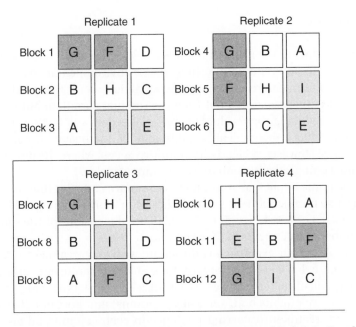

Fig. 11.7 *Lattice design of incomplete blocks:* Here pairs in a block (e.g. G and F or E and I in Block 1) do not coincide in another block. Replicate 2 is replicate 1 with the blocks now used as columns. Replicate 3 uses the G-H-E diagonal as the first block, then B-I-D from a second diagonal of replicate 1 assuming a continuously repeating pattern, and the third block is the next diagonal (A-F-C of block 1). The blocks of replicate 4 use diagonals from replicate 3.

a block in replicate 3. In the plan below, the pairs G and F, also I and E, are shaded to illustrate the principle of nonrepeating pairings in the different blocks.

Note: *Had the fertilizer treatment been three increasing levels rather than three different fertilizers, there is a more advanced approach to analyzing the data based on the presumption that the means will follow (either positively or negatively) the increase in fertilizer applied. This is outside the scope of this book, but details can be found in more advanced texts. The approach is to calculate what proportion of the treatment variability (as sum of squares) can be apportioned to single degrees of freedom associated with fitting linear or defined nonlinear functions to the sequence of means. These functions are mentioned briefly on page 269 in Chapter 17. The advantage of this different approach is that, if the means do follow such directional changes the degrees of freedom used to calculate the mean square are reduced and there is therefore a greater chance of detecting the effect of fertilizer as statistically significant.*

Latin square

When introducing the randomized block design, I pointed out that it was appropriate to an arena with a single major environmental or other gradient. The Latin square design enables variation along two gradients at right angles to be separated out from the residual variation, but with the penalty of reducing the residual degrees of freedom still further from the randomized block design. This penalty may be too heavy for comfort in many cases! However, it clearly has intuitive appeal where the direction of any important gradients is unknown or unpredictable.

As the name implies, a Latin square is a square design in that it consists of the same number of plots (though these don't have to be square) in two dimensions, i.e. 4×4, 5×5, 6×6, etc. The "dimension" of the square is the number of treatments in the experiment.

We'll take as our example a Latin square with four treatments (A, B, C, and D) – i.e. a 4×4 (16 plot) square. To be able to separate out variation in the two gradients at right angles, we visualize the 16 plots as being "blocks" of the four treatments both vertically (columns) and horizontally (rows), with each treatment occurring just once in each column and also only once in each row. Apart from that restriction, treatments can be allocated to plots at random. The layout might therefore be as in Fig. 11.8a (the process of randomization is described in Box 11.4): Note that the design results in "blocks" in two directions (Fig. 11.8b).

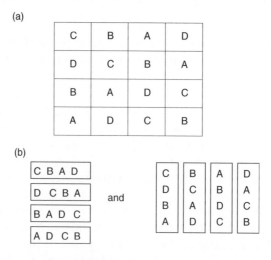

Fig. 11.8 *Latin square design:* (a) A possible layout for four treatments; (b) how these could be either vertical or horizontal blocks of a randomized block design – in either case, each treatment would appear once in each block.

BOX 11.4

We begin with a randomization in one direction:

C   B   A   D

Since C occupies the top left corner, we just have A, B, and D to randomize for the vertical dimension:

C   B   A   D
D
B
A

The next row (following D) leaves A, B, and C to randomize, with the constraint that the letter must be different from the one above it in the first row: So a randomization of C, A, and B had to be changed to C B A

C   B   A   D
D   C   B   A
B
A

The next row beginning with B involves randomizing A, C, and D, again so that no letter in a column is repeated. What came up was A D C, which is fine

C   B   A   D
D   C   B   A
B   A   D   C
A

The final row completes the missing letter in each column

C   B   A   D
D   C   B   A
B   A   D   C
A   D   C   B

Data for the analysis of a Latin square

The arrival of winged aphids on a crop is usually very directional, but the direction varies from season to season. A Latin square design was therefore deemed appropriate for a trial on the suitability of four varieties of cabbage

C 69	B 58	A 73	D 91
D 114	C 59	B 44	A 52
B 34	A 46	D 75	C 18
A 34	D 62	C 11	B 4

Fig. 11.9   *Latin square design:* The layout of Fig. 11.8a with data added. The experiment measured the number of winged migrants of cabbage aphid remaining on four cabbage varieties the morning after a migration which most years comes in at the top left corner of the field (see text).

to arriving cabbage aphids (aphids tend to alight equally on the varieties in the evening, but show their preference by whether or not they take off again the next morning). The data (Fig. 11.9) are therefore the total number of winged cabbage aphids found on 20 plants of four cabbage varieties (A, B, C and D) at 11 a.m. after water dishes among the plants had shown an arrival flight had occurred.

Prelims

We will need column and row totals – the tabulated numbers to the left below (check the grand total both ways from the end totals); but we will also need to unscramble the randomization to obtain the treatment end totals – the tabulated numbers on the right. Note again the use of subscripts with the totals we will need to SqADS, to denote how many data have been totaled (remember, this subscript becomes the divisor of the added squared totals in SqADS).

						A	B	C	D
69	58	73	91	291_4		73	58	69	91
114	59	44	52	269_4		52	44	59	114
34	46	75	18	173_4		46	34	18	75
34	62	11	4	111_4		34	4	11	62
251_4	225_4	203_4	165_4	GT = $\mathbf{844}_{16}$		205_4	140_4	157_4	342_4	GT = $\mathbf{844}_{16}$

Note: The end totals of the original data show a diagonal decline from the top left corner, suggesting that the aphid flight came from this direction.

What is the lead line? It cannot be "4 varieties × 4 columns × 4 rows = 16 plots," since $4 \times 4 \times 4 = 64$, not 16! With a Latin square design, there is clearly one "×4" too many! We are used to analyzing a table like the one on the left above (4 columns ×4 rows = 16 plots) according to the table below (shown in small type only, because it is **NOT** the final table appropriate for a Latin square design):

Lead line: 4 Columns × 4 Rows = 16 plots

Source of variation	Degrees of freedom	Sum of squares	Mean square	Variance ratio	P
4 Columns	3				
4 Rows	3				
Residual	$(3 \times 3 =) 9$				
TOTAL (16 plots)	15				

and the degrees of freedom add up OK. But where are the Varieties in the above skeleton analysis table? They are there, I promise you. Believe it or not, they are in the Residual! Look back at the section on randomized Blocks and particularly Box 11.2. Here the point was made that the residual represented the inconsistency (or interaction) between Treatments and Blocks (= Columns and Rows in a Latin square), i.e. the failure of plots to yield data consistent with the overall effect of their Treatment and Block totals (= Column and Row). Now that's exactly what the Varieties are doing. They are adding a Variety "weighting" to the random residual variation of each plot. So both the degrees of freedom and sum of squares for the variation caused by having four different Varieties in our Latin square have to be separated out from the residual, which therefore loses 3 degrees of freedom for the four Varieties.

The correct analysis of variance table is therefore:

Lead line: 4 Columns × 4 Rows = 16 plots of 4 Varieties

Source of variation	Degrees of freedom	Sum of squares	Mean square	Variance ratio	P
4 Varieties	3				
4 Columns	3				
4 Rows	3				
Residual	$(3 \times 3 - 3 =)$ 6				
TOTAL (16 plots)	15				

Phase 1

The Correction Factor (Grand Total$^2/n$) is $844^2/16 = 44,521$
Added squares in the body of the table are $69^2 + 58^2 + 73^2 + \cdots + 4^2 = 57,210$
Therefore Total Sum of Squares $= 57,210 - \text{CF} = 12,689$

Phase 2

This phase uses only end totals (the totals with subscripts in the two adjacent tables on page 133) in the calculations). For each systematic source of variation (Varieties, Columns, and Rows in our example here) we use the SqADS procedure on the appropriate totals from the two tables, remembering that the divisor (D) is the subscript indicating how many data contribute to the total, and that the number of totals we square and add (SqA) multiplied by the divisor $=$ the total number of plots (16 in our example).

So sum of squares for Varieties $= \mathbf{SqA}^{4 \text{ Variety totals}} \mathbf{D}^{\text{by } 4} \mathbf{S}^{\text{Correction Factor}} = 6288.50$

Sum of squares for Columns $= \mathbf{SqA}^{4 \text{ Column totals}} \mathbf{D}^{\text{by } 4} \mathbf{S}^{\text{Correction Factor}} = 994.00$

and sum of squares for Rows $= \mathbf{SqA}^{4 \text{ Row totals}} \mathbf{D}^{\text{by } 4} \mathbf{S}^{\text{Correction Factor}} = 5302.00$

Once again, for those who prefer it, the calculations are repeated with actual numbers in Box 11.5.

BOX 11.5

Phase 2 sum of squares calculations using the Variety, Column, and Row totals from the two adjacent tables on page 133, together with the correction factor already calculated in Phase 1.

$$\text{Varieties sum of squares} = \frac{205^2 + 140^2 + 157^2 + 342^2}{4} - 44521 = 6288.50$$

$$\text{Columns sum of squares} = \frac{251^2 + 225^2 + 203^2 + 165^2}{4} - 44521 = 994.00$$

$$\text{Rows sum of squares} = \frac{291^2 + 269^2 + 173^2 + 111^2}{4} - 44521 = 5302.00$$

Note that in both cases the number of totals squared and added $\times$ the divisor $= 16$, the number of plots in the experiment.

These calculations are inserted into the analysis table:

Lead line: 4 Columns × 4 Rows = 16 plots of 4 Varieties
Correction Factor: 44,521

Source of variation	Degrees of freedom	Sum of squares	Mean square	Variance ratio	P
4 Varieties	3	6288.50			
4 Columns	3	994.00			
4 Rows	3	5302.00			
Residual	(3 × 3 − 3 =) 6	104.50			
TOTAL (16 plots)	15	12,689.00			

End Phase

The last phase is to calculate mean squares by dividing the sums of squares by their degrees of freedom and then dividing all the other mean squares by the residual mean square to give the variance ratios. The phase concludes with checking tables of F for the P (probability value) of the variance ratios. The value in the table will be the same for all the variance ratios calculated, since in each case it will be at the intersection of 3 d.f. across the top of the F table and 6 Residual d.f. down the side:

Lead line: 4 Columns × 4 Rows = 16 plots of 4 Varieties
Correction Factor: 44,521

Source of variation	Degrees of freedom	Sum of squares	Mean square	Variance ratio	P
4 Varieties	3	→ 6,288.50	→ 2096.17	120.35	<0.001
4 Columns	3	→ 994.00	→ 331.33	19.02	<0.01
4 Rows	3	→ 5,302.00	→ 1767.33	101.47	<0.001
Residual	(3 × 3 − 3 =) 6	104.50	17.42		
TOTAL (16 plots)	15	12,689.00			

All three variance ratios are so large that the chance of them being calculated in the absence of variation from the sources involved is less than once in 1000 for Varieties and Rows and once in 100 for Columns. The Latin square design has succeeded in keeping the residual variation restricted to background interplot variability and has separated out both the strong column and row gradients associated with the directional arrival of the aphids from the top left hand corner of the field (page 133). The variance ratio for Varieties would have been reduced if it had not been laid out as a Latin square, since either the Column or Row sum of squares would have been included in the residual variation (though with increased degrees of freedom). The relevant analysis tables are shown in

BOX 11.6

Analysis if Columns had been the randomized blocks:

Source of variation	Degrees of freedom	Sum of squares	Mean square	Variance ratio	P
4 Varieties	3	6288.50	2096.17	0.35	n.s
4 Columns	3	994.00	331.30	0.06	n.s
Residual	(3 × 3 =)9	54,065.50	6007.28		
TOTAL (16 plots)	15	12,689.00			

Varieties would have shown a negligible variance ratio of only 0.35, likely to arise very commonly by chance in the absence of any Variety effect.

Analysis if Rows had been the randomized blocks:

Source of variation	Degrees of freedom	Sum of squares	Mean square	Variance ratio	P
4 Varieties	3	6288.50	2096.17	17.17	<0.01
4 Rows	3	5302.00	1767.33	14.48	<0.01
Residual	(3 × 3 =)9	1098.50	122.05		
TOTAL (16 plots)	15	12,689.00			

The Variance ratio for Varieties would still have been significantly high, but with the probability of it occurring by chance increased. It is also possible that statistical discrimination between the varieties might have been reduced (see Chapter 16).

Box 11.6 – showing a reduction in F for Varieties from the 120.35 above to 17.17 if Rows had been the blocks and only 0.35 if it had been the Columns.

The further tests on the Variety means needed to interpret the experiment fully will be presented later in Chapter 16, page 229.

Further comments on the Latin square design

Since Latin squares use up so many degrees of freedom for systematic variation (Treatments, Rows, and also Columns) at the expense of degrees of freedom for the residual, they are rather inefficient at measuring the latter accurately. However, it is clear from our aphid example above that they have great value if a strong gradient is expected but its direction is

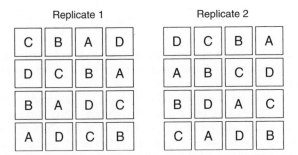

Fig. 11.10 *Latin square designs*: A possible randomization of four treatments in two replicate 4 × 4 Latin squares. Each square has a fresh independent randomization.

variable or unknown. That gradients are often unknown makes it profitable to try a Latin square as a pilot experiment in a new environment. The results will make it clear whether a Latin square is necessary or whether there is just one gradient which necessitates blocking in the right direction. Of course, if no gradient is detectable, then a fully randomized experiment may well be the best policy following the pilot experiment.

However, what can be done to increase the residual degrees of freedom without adding more treatments if it appears a Latin square is the right design for the situation? As demonstrated in the aphid example, experiments with wind-dispersing pests, diseases, or weeds are all likely candidates. The answer is that residual degrees of freedom can be increased easily by setting up more than one replicate Latin square (with different randomizations of course) as in Fig. 11.10.

How such multiple Latin squares are analyzed is a bit advanced for a book at this level, but I wasn't able to find out how to do it in any of the larger higher-level textbooks I consulted. Having worked out how to do it, I realized that the allocation of degrees of freedom and the calculation of some of the sums of squares are a little idiosyncratic. I have therefore provided guidelines in Box 11.7, though you are probably best advised to skip over this until the day comes when you need it!

Split plot

This design is added here just as a heading (for completing the range of common experimental designs). It has some resonance of the multiple Latin square, in that an experiment of large plots (like replicate entire Latin squares) has another experiment of smaller plots going on inside it (like the columns within each replicate Latin square).

BOX 11.7

For the allocation of degrees of freedom and calculation of sums of squares for a multiple Latin square, we'll use the example of two 4 × 4 squares illustrated in Fig. 11.10 on page 137.

This is the one type of analysis of variance where we depart from the rule stated on page 93 that we use the **same** correction factor **throughout**! This complication arises with the sums of squares for the Columns and Rows sources of positional variation, since some of this variation is the same (positional) variation as already included in the sum of squares for variation shown by the totals of the two replicate Latin squares (* in the table below).

We deal with this by calculating a pooled sum of squares as for the t-test (page 68), where we work out the sum of squares for each replicate square separately (using the appropriate replicate square total for the Correction Factor, **not** the combined Grand Total (GT)). Similarly we also pool the two separate degrees of freedom for the two squares as the divisor for calculating the mean squares for Columns and Rows:

Source of variation	Degrees of freedom	Sum of squares
4 Treatments (A–D)	3	$Sq_A^{4\ \text{Treatment totals}}{}_D$by 4 $S^{GT\text{squared}/32}$
*2 Replicate squares	1	$Sq_A^{2\ \text{Square totals}}{}_D$by 16 $S^{GT\text{squared}/32}$
Twice 4 Rows	(2 × 4 − 1) 6	$Sq_A^{4\ \text{Rep.1 Row totals}}{}_D$by 4 $S^{\text{Rep.1 total(squared)}/16}$ **plus** $Sq_A^{4\ \text{Rep.2 Row totals}}{}_D$by 4 $S^{\text{Rep.2 total(squared)}/16}$
Twice 4 Columns	(2 × 4 − 1) 6	$Sq_A^{4\ \text{Rep.1 Row totals}}{}_D$by 4 $S^{\text{Rep.1 total(squared)}/16}$ **plus** $Sq_A^{4\ \text{Rep.2 Row totals}}{}_D$by 4 $S^{\text{Rep.2 total (squared)}/16}$
Residual	by subtraction 15	by subtraction
TOTAL (32 plots)	31	$Sq_A^{32\ \text{data}}{}_S{}^{GT\text{squared}/32}$

It is therefore a design for more than a single series of treatments and appropriate only for factorial experiments (see Chapters 12–14) where the "treatments" are combinations of at least two factors (e.g. six "treatments" might represent two crop types, each given one of three levels of fertilizer).

The split plot design will be allocated its own chapter (Chapter 15), after "nonsplit" factorial experiments have been described.

EXECUTIVE SUMMARY 5
Analysis of a randomized block experiment

If necessary, convert the data to a table with the randomization of treatments "unscrambled."

Phase 1

1 Total rows and columns, and check Grand total by adding both row and column totals.

2 Calculate correction factor for experiment $=$ Grand total^2/number plots.

3 Calculate **TOTAL** sum of squares (added squares − **correction factor!**).

YOU HAVE NOW NO FURTHER USE FOR THE PLOT RESULT DATA

Phase 2

4 Construct the analysis of variance table with the headings: *Source of variation, d.f., Sum of squares, Mean square, F, and P.*

5 Insert sources of variation as *Treatments, Blocks, Residual,* and *Total.*

6 Allocate degrees of freedom as $n - 1$ for number of Treatments and Replicates, and the product of these two d.f. for the Residual. Check that the three d.f.'s add up to $n - 1$ for the Total (i.e. one less than the number of data in the experiment).

7 **SqADS** the TREATMENT totals to obtain the *Treatments* sum of squares$_{\text{of deviations}}$.

8 **SqADS** the REPLICATE totals to obtain the *Replicates* sum of squares$_{\text{of deviations}}$.

9 The "Residual" sum of squares$_{\text{of deviations}}$ is the **remainder** of the "total" sum of squares$_{\text{of deviations}}$.

YOU HAVE NOW USED ALL THE COLUMN AND ROW TOTALS

End Phase

10 Calculate *mean square* for "treatments," "replicates," and "residual" by dividing each sum of squares$_{\text{of deviations}}$ by its own degrees of freedom.

11 Calculate F for "treatments" and "replicates" by dividing their mean squares by that for the residual.

12 Check P (= probability of finding as large an F by chance) for "treatments" and "replicates" in tables of F values.

s.e.d.m. = standard error of difference between 2 means = $\sqrt{2s^2/n_{\text{of the mean}}}$. The t-test calculates whether a difference (between two means) divided by the s.e.d.m. is greater than the tabulated value of t (for $P = 0.05$ and for the "residual" degrees of freedom).

This is the same as accepting any difference between two means greater than $t \times$ s.e.d.m. as statistically significant. "$t \times$ s.e.d.m." is therefore called the "LEAST SIGNIFICANT DIFFERENCE."

s^2 for calculating $\sqrt{2s^2/n_{\text{of the mean}}}$ for the experiment after an analysis of variance is the "residual mean square." Double it, then divide by the number of plots contributing to each treatment mean. Square root the result, and this is your s.e.d.m. to multiply by "t" for "least significant difference" tests (but see Chapter 16).

Spare-time activities

1 The data are yields (kg dry matter/plot) of sweet corn grown at five different plant densities in five replicates.

Replicates	Plants/m^2				
	20	**25**	**30**	**35**	**40**
1	21.1	26.8	30.4	28.4	27.6
2	16.7	23.8	25.5	28.2	24.5
3	14.9	21.4	27.1	25.3	26.5
4	15.5	22.6	26.3	26.5	27.0
5	19.7	23.6	26.6	32.6	30.1

What design is this? Complete the appropriate analysis of variance.

Use the LSD (page 109) for differences between adjacent means to identify which increases in spacing significantly change the yield.

2 In a comparison of four wild species of potato (A, B, C, and D), one of the characters the plant breeder measured was the number of insect repellent glandular hairs/mm^2 on 10 plants of each species:

A	B	C	D
1.1	3.7	4.9	6.4
1.8	4.8	9.4	15.2
1.0	1.6	2.9	10.5
1.5	3.1	5.4	11.2
1.3	5.1	3.8	8.4
0.0	5.2	6.1	6.5
1.7	7.5	5.9	7.4
1.3	6.5	5.4	14.6
1.6	6.7	6.8	7.8
1.7	6.9	5.9	11.1

What design is this? Complete the appropriate analysis of variance.

Use the LSD test (page 109) to identify which varieties have a hair-density significantly different from the normal commercial variety, which is variety B.

3 Six lettuce varieties (identified in the table below by code letters) were grown in small plots by each of eight students. Lettuces put on most of their fresh weight in a short period just before maturity, so all the data were taken at harvest of the varieties for marketing, and not on the same calendar date. Even so, variances of the different varieties were very unequal, and so the data were transformed to log$_{10}$ before statistical analysis (page 38).

The data are therefore the logarithm$_{10}$ of the mean weight in grams per plant based on five or six plants taken from each plot.

Student	\multicolumn{6}{c}{Lettuce variety}
	B	D	AV	G	L	V
J. Blogg	2.55	2.48	2.74	3.01	2.79	2.93
P. Ellis	2.39	2.41	2.60	2.92	2.63	2.85
A. Norris	2.44	2.37	2.64	2.86	2.69	3.01
F. Newton	2.46	2.43	2.73	2.93	2.79	2.95
J. Skeete	2.55	2.44	2.66	2.86	2.81	2.92
D. Taylor	2.51	2.39	2.74	2.98	2.67	2.92
J. Taylor	2.50	2.51	2.77	3.01	2.80	2.80
A. Tinsley	2.54	2.46	2.71	2.95	2.70	2.93

Is there good evidence that students can have a bigger effect on lettuce yields than seed packets?

4 Two growth promoters coded P-1049 and P-2711 were applied to *Zinnia* plants, and the day from sowing to flowering of the individual plants was recorded. Unfortunately, P-1049 showed some toxicity to the seedlings and some of the plants died before flowering.

No treatment	P-1049	P-2711
64	107	70
78	74	74
93	83	86
80	65	79
89	72	84
79	68	75
91	69	87
102	68	99
71		70
106		83

Do either of the growth promoters also speed up flowering? Use the LSD test (page 109) if the variance ratio for treatments is significant at $P = 0.05$.

12

Introduction to factorial experiments

Chapter features

What is a factorial experiment?

In Chapter 11, I described a number of experimental designs in which each "plot" (e.g. part of a field or a Petri dish in an incubator) was assigned to one of the experimental treatments. Such treatments were variations of a single *factor* – different varieties of crop plant, different fertilizers, different application rates of the same fertilizer, etc. These variations of one factor are called "*levels.*"

A *factorial experiment* is one where the treatments allocated to the experimental plots are **combinations** of two or more factors (hence the term "factorial"). For example, we might wish to test the effect on some measurement of a NUMBER of different fertilizers on MORE THAN ONE plant variety. Each plot would then be allocated the combination of one "level" of fertilizer (one of the fertilizers) applied to one "level" of variety (one of the varieties).

So we'll use the above experiment as our example, using three fertilizers (A, B, and C) applied to two plant varieties (Y and Z). Three levels of fertilizer × two levels of variety = six combinations, which are AY, BY, CY, AZ, BZ, and CZ. A single replicate of the experiment thus requires six plots, one for each "treatment," with each treatment being a different combination of levels of the two factors. These replicates of six treatments can be laid out as any of the designs described in Chapter 11, and Fig. 12.1 shows these same six treatments laid out as fully randomized, randomized block, and a Latin square designs.

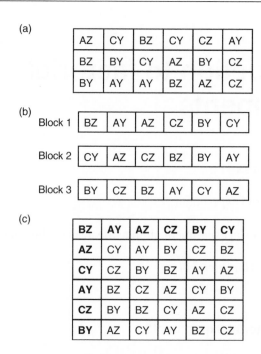

Fig. 12.1   The same six factorial combinations of two factors laid out as: (a) a fully randomized design of three plots per treatment; (b) three randomized blocks; (c) a 6 × 6 Latin square.

Instead of applying three fertilizers to our two varieties (Y and Z), we might wish to apply just one fertilizer, but at THREE rates (**H**igh, **M**edium, and **L**ow). This is still a 3 level × 2 level factorial, again with six combinations (or plots per replicate), viz. YH, YM, YL, ZH, ZM, and ZL.

Another 3 level × 2 level factorial might again involve one fertilizer but at the three rates (H, M, and L) applied at two different times (**E**arly and **D**elayed) in the season. This would again give six "treatments": HE, HD, ME, MD, LE, LD.

We might even get more complicated still, and bring back our two plant varieties (Y and Z) into the last (rates × timings) factorial, now giving us a three-factor experiment of 3 fertilizer levels × 2 fertilizer timings × 2 varieties = 12 "treatments": HEY, HEZ, HDY, HDZ, MEY, MEZ, MDY, MDZ, LEY, LEZ, LDY, and LDZ.

Note: When the data come from increasing or decreasing levels of the same factor (e.g. Low, Medium, and High fertilizer levels in the above examples), the "note" on page 129 applies. In this chapter we shall treat these three levels of fertilizer as if they were not in an ascending series and could equally have been three different fertilizers. So we will calculate a sum of squares for these three levels of fertilizer to 2 degrees of freedom. But be aware of the possibility outside the scope of this book but mentioned on page 130 of increasing the F value by fitting linear or nonlinear functions to the sequence of means.

Interaction

I have found that the concept of *interaction* is a real hurdle for biologists. Strangely it is not the computational procedures which seem to be the problem, but the explanation of the interaction in plain English, i.e. what "biological" phenomenon the interaction reveals! Before looking at interaction in biological experiments, I want to use a rather silly set of data, the time taken for a person walking, a cyclist, and a car to cover three different distances (a 3 transport levels × 3 distance levels factorial = 9 data). The table of data is shown in Fig. 12.2a, with only the end totals of rows (distances) and columns (modes of transport) provided.

Rather predictably, the totals for the three levels of the transport factor show the longest total time over the three distances for the walker and the shortest for the car, with the bike in the middle! Surprise, surprise, the total times over the three transport levels show it takes longer to travel 2000 m than 500 m, with 1000 m in the middle.

The possibility of *interaction* arises because an identical set of end totals can arise from many different sets of data in the body of the table.

If there is no interaction, then the two factors are acting independently in an additive way. The concepts of "additivity" and "nonadditivity" are frequently encountered when shopping! A certain computer may be £700 and a particular model of digital camera might be £250. How much would it cost to buy both? Well, with no special offer, the prices would be "additive" and we would have to fork out £950. But there might be a special offer (=nonadditivity) of the job lot for only £800! Now let's go back to our "mode of transport × distance" example of a factorial.

If there is no interaction

No interaction between the two factors would mean that the time increases shown by the row totals with longer distances apply equally to all modes of transport. So the bike would take 1.26 times longer than the car at all distances, and the walker 8 times longer than the bike. Equally all modes of

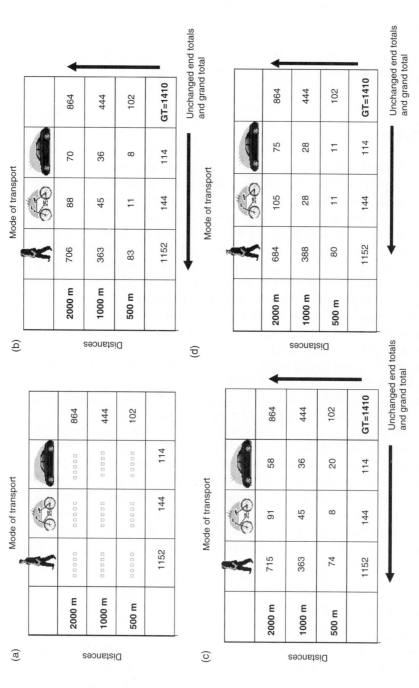

Fig. 12.2 Time in seconds taken by a walker, a bike, and a car to cover 500, 1000, and 2000 m. (a) The row and column totals without the data in the body of the table; (b–d) three different sets of data which add up to the same row and column totals as in (a). (b) Data with no interaction between mode of transport and distance and (c,d) two different data sets, both showing interaction (see text).

transport would take the same roughly four times longer to travel 1000 m than 500 m and then double the 1000 m time to cover 2000 m.

There is only one set of numbers for the body of the table which would fit these criteria – after all, there is no further information to add that is not already provided by the row and column totals. This unique set of numbers is shown in Fig. 12.2b. How did I calculate them? Well, this was explained in Box 11.3, where I tackled the identical exercise of calculating plot data from the replicate (row) and treatment (column) end totals with the assumption that the treatments behaved identically relative to each other in all replicates. Check out for yourself that the numbers in Fig. 12.2b show the same incremental steps as the end totals.

What if there **is** interaction?

Such consistent relationships between the modes of transport may not hold if the traffic is very heavy, and it is possible that a short distance might then be faster by bike than by car. Figure 12.2c shows this situation at 500 m, where the bike has halved the time it takes by car. The rows and columns still add up to the same end totals with the **overall** time taken by the bicycle over all three distances still just as much more as before than the time taken by the car in spite of its shorter time at the one distance of 500 m.

The ranking of the three modes of transport at the three distances has become inconsistent. This inconsistency is an *interaction* – and here represents the "transport × distance interaction." It would clearly be misleading to rely on end totals and state "the car takes less time than the cycle," since at 500 m this is palpably untrue!

A different interaction, yet still with the same row and column end totals (Fig. 12.2d)

Another way the interaction might show is when the roads are even busier, so that the cycle is slowed by the heavy traffic as much as the car for distances up to 1000 m.

It would again be wrong to base our conclusions on the end totals, and claim that the car is faster than the bike, because such a claim is only true for a distance of 2000 m; there is no difference at all for 1000 or 500 m.

Recap on Fig. 12.2

All four tables in Fig. 12.2 share the same end totals, which therefore indicate the same differences in all the tables in (i) the speed between the modes of transport and (ii) the times taken to cover the three distances.

However, **that** interpretation is only valid in respect of the Fig. 12.2b. In Fig. 12.2c that interpretation is **not** valid, because over 500 m the bike is faster than the car and in Fig. 12.2d there is no difference between these two modes of transport, except at just one (the longest) distance.

Figures 12.2c and 12.2d are examples of *interaction* between the two factors (mode of transport and distance) which cannot be predicted from the row and column end totals, as they could be if interaction is not occurring (Fig. 12.2b).

How about a biological example?

For those who would find a second example of interaction helpful, a table of plant yields in a factorial experiment of three fertilizers applied to two crop varieties is given in Box 12.1. I suggest you look at the numbers only, and try and deduce for yourself whether there is interaction and how you would explain any interaction you spot in biological terms.

Measuring any interaction between factors is often the main/only purpose of an experiment

To make the point, I can recall an experiment in my research group which looked at the resistance to a pest (the diamond-back moth) of four crop *Brassica* species. But we were already pretty sure of the resistance ranking of the brassicas from published glasshouse experiments, and also knew (from previous work in the group) that plants grown outdoors have higher concentrations of chemicals which are distasteful or toxic to plant-feeding insects than plants grown in the glasshouse. So, when we tested the success of diamond-back moth caterpillars feeding in the lab on leaves of the four brassicas (factor 1) grown under both outdoor and glasshouse conditions (factor 2), we were already pretty certain what the overall effect of brassicas and of location would be (i.e. even before the experiment, we could predict the row and column end totals of a table such as Fig. 12.2a). The only reason for doing the experiment was what the numbers in the body of the table would be! Would they show *interaction* between the two factors. That is, would the order of resistance of the four brassicas be different in the glasshouse from what it was outdoors? Of course, it was!

Very often, at least one of the factors in a factorial experiment is only included because of the interest in what interaction it might have with another, yet so often I"ve seen student project reports describing the results of a factorial experiment only with reference to the "main effects" of the factors (i.e. only the column and row end totals have been discussed and

BOX 12.1

For this example of potential interaction, here are possible data from the first 3×2 level factorial mentioned in this chapter – 3 fertilizers (A, B, and C) applied to two crop varieties (Y and Z). The data are yield (kg/plot), averaged across replicates.

First data set

	A	B	C	Mean
Y	37	39	29	**35.0**
Z	25	27	18	**23.3**
Mean	**31.0**	**33.0**	**23.5**	

Interaction: Very little. Variety Y outyields variety Z with all three fertilizers, though to varying degrees (42, 44, and 55%).
Variety overall: Y consistently outyields Z.
Fertilizer overall: Yields with A and B are very similar and higher than with C.

Second data set

	A	B	C	Mean
Y	37	39	20	**32.0**
Z	25	27	35	**29.0**
Mean	**31.0**	**33.0**	**27.5**	

Interaction: Important. The effect of fertilizer C on the varieties (Z outyields Y) is the reverse of A and B.
Variety overall: Y outyields Z – **WRONG!** Does not apply to fertilizer C.
Fertilizer overall: Yields with A and B are very similar and higher than with C. – **WRONG!** With variety Z, C gives a yield higher than A or B.

Third data set

	A	B	C	Mean
Y	37	39	20	**32.0**
Z	25	27	25	**29.0**
Mean	**31.0**	**33.0**	**23.5**	

Interaction: Important. Z does not seem to respond differently to the 3 fertilizers.
Variety overall: Y outyields Z – **WRONG!** Does not apply to fertilizer C.
Fertilizer overall: Yields with A and B are very similar and higher than with C. – **WRONG!** This only applies to variety Y; Z shows no clear differences.

compared). The most interesting result of the experiment is then summarily dismissed with the brief phrase "The interaction was also significant ($P < 0.001$)." Of course, there's no "also" about it – IF the interaction is statistically significant then that it is very likely the most important result of the experiment, and certainly should be reported properly in biological (and not merely statistical) terms. Moreover, great care needs to be taken in making any statements, if any at all are worth making, about the ranking of the end totals. This is because such totals average what may be inconsistent rankings of means across treatments, as in the earlier "transport" example where which mode of transport is fastest cannot be deduced from the end totals; it all depends on which distance is involved!

How does a factorial experiment change the form of the analysis of variance?

Degrees of freedom for interactions

We'll use the example given earlier in this chapter and in Fig. 12.1b of 3 levels of fertilizer $\times$ 2 levels of plant variety ($=6$ combinations or "treatments") laid out in 3 randomized blocks (replicates). There would be $6 \times 3 = 18$ plots in the experiment.

There is no change to Phase 2 of the analysis of variance (see page 118) – we still have 6 "treatments" and 3 replicates with degrees of freedom as:

Source of variation	Degrees of freedom
6 Treatments	$5(=6-1)$
3 Replicates	$2(=4-1)$
Residual	$10(5 \times 2 \text{ or } 17-(5+2))$
Total (18 plots)	17

The "treatments" in this experiment are now combinations of levels of two factors, so it is possible for there to be *interaction* between them. The treatment variation is therefore the sum of **three** different sources of variation: the fertilizer factor, the variety factor, and the possible interaction.

For sorting out how much variation to ascribe to each of these three components of the variation due to the six "treatments," the analysis of a factorial experiment has to include a new Phase 3 inserted before the

End Phase of Anova. Just as Phase 2 partitions the sum of squares for the "Total" variation into three components (treatments, replicates, and residual) so Phase 3 of a two-factor factorial partitions the "Treatments" into factor 1, factor 2, and their interaction as (just check the similarity of thinking):

	Source of variation	Degrees of freedom
Phase 3	3 Fertilizers	$2(=3-1)$
	2 Varieties	$1(=2-1)$
	Interaction	$2(2 \times 1$ or $5-(2+1))$
From Phase 2	6 Treatments	5

The way d.f. for Fertilizers and Varieties are allocated should by now be familiar – but isn't there also something familiar about the d.f. for the interaction? Just as for the residual in Phase 2, we **multiply the degrees of freedom for the two components of the interaction** (in this case the fertilizer $\times$ variety interaction). Is this a coincidence?

The similarity between the "residual" in Phase 2 and the "interaction" in Phase 3

How long ago did you read page 124? Perhaps you should now read it again before going on with this chapter. The interaction is a **residual part** of the total "treatment" variation. Indeed, note above that we find the d.f. by subtracting the d.f. for the main factor effects from the d.f. for treatments. Equally, the residual in Phase 2 is an **interaction** (between treatments and replicates).

Earlier in this chapter I tried to ram home what interaction means – that it is the degree to which the data for combinations of factor levels differ from what one would predict from the effect of the factor levels in isolation (the row and column totals or means (as in Box 12.1)).

Another way of putting this is: how inconsistent are the effects of factor 1 at different levels of factor 2? In just the same way: how inconsistent are the effects of the treatments in the different replicates? Both "interaction" and the "residual" variation measure the degree to which the plot data cannot be predicted from the row and column end totals. I hope the similarity of interactions in our new Phase 3 with the residual in Phase 2 is now clear? Perhaps Box 12.2 can help.

BOX 12.2

Equivalence of the interaction (in italics) and residual (in bold) sums of squares can be illustrated in a table:

Source of variation	Degrees of freedom	Sum of squares
Factor 1 or **Treatments**	a	Can be computed from the data
Factor 2 or **Replicates**	b	Can be computed from the data
Interaction or **Residual**	a × b	Can **ONLY** be found as a **remainder** (i.e.) by subtraction
Treatments or **TOTAL**	a + b + (a × b)	Can be computed from the data

Sums of squares for interactions

If the philosophy of an interaction in Phase 3 is akin to that of the residual in Phase 2, we already know how to calculate its sum of squares. **It will be a remainder found by subtraction** (see Box 12.2). It will be what is left of the treatment sum of squares from Phase 2 when we have subtracted the sums of squares calculated for the two factors (Fertilizer and Variety in our example).

The big difference between residual and interaction sums of squares lies in the purpose behind calculating them! The interaction sum of squares measures variability which is part of the systematic design of the experiment and therefore it has its mean square and variance ratio (see page 162) tested in just the same way as the individual factors that may be interacting. The residual sum of squares has its mean square calculated as the yardstick with which we compare the other mean squares (as variance ratios), **including** the mean square for interaction. The latter is NOT some other measure of "experimental error." The way people ignore interactions often makes me wonder whether they think it is!

Like the residual sum of squares, interaction sums of squares are always calculated by subtraction, i.e. as **remainders**. I find this is something that people tend to forget, so do try and remember it! To calculate the interaction sum of squares (described in the next chapter) we have first to **SqADS** the totals representing the combined variation due to the interaction **and** the factors that are interacting. After **S**$^{\text{correction factor}}$ we haven't finished. We still have to subtract the sums of squares for the factors that are interacting to find that elusive remainder which measures the variability caused by the interaction.

So it can't be repeated too often – INTERACTION SUMS OF SQUARES ARE ALWAYS **REMAINDERS**.

13

2-Factor factorial experiments

Chapter features

Introduction

With factorial experiments, the value of having the *lead line* and using *subscripts* to track the number of data contributing to totals will become more evident. But this chapter also includes three new ideas, the introduction of a *Phase 3* into the analysis (already introduced in Chapter 12), the combining of "treatment" totals into *supertotals* and adding a column to the Anova table for the divisor appropriate to each SqA**DS** calculation for sums of squares. As in previous chapters, some new ideas introduced may seem unnecessary, but they will prove their worth in more complex factorial experiments (Chapter 14) and are best introduced in a simpler situation while they still appear "easy."

An example of a 2-factor experiment

We'll add some data (Fig. 13.1) to the example of a randomized block design of the 2-factor experiment with which we introduced Chapter 12. The data are yield (kg/plot) of two strawberry varieties (Y and Z) treated with three different fertilizers (A, B, and C).

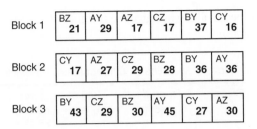

Fig. 13.1 A 2-factor randomized block experiment with three blocks applying three different fertilizers (A, B, and C) to two strawberry varieties (Y and Z). Data are fruit yield as kg/plot.

Analysis of the 2-factor experiment

We can begin with what is hopefully by now very familiar territory and treat this like any other randomized block experiment with six treatments and three replicates – through *prelims* and *Phase 1* right up to the end phase of *Phase 2* before drawing breath!

Prelims

Unscramble the randomization so that each plot of each of the six treatments is in the same column (or row if you prefer). Note the subscripts that have again been added to the end totals to remind us how many data contribute to each total.

	AY	BY	CY	AZ	BZ	CZ	
Block 1	29	37	16	17	21	17	137_6
Block 2	36	36	17	27	28	29	173_6
Block 3	45	43	27	30	30	29	204_6
	110_3	116_3	60_3	74_3	79_3	75_3	GT = $\mathbf{514_{18}}$

The row (block) totals suggest a gradient of increasing yield down the table; the correct alignment of the blocks appears to have been used (page 122).

The lead line is added and used to produce the skeleton analysis table (at this stage just for the six "treatments" ignoring their factorial nature) with degrees of freedom $(n - 1)$. Note the new column headed "Divisor," this provides the D in SqADS and, when multiplied by the number of totals to be

Squared and **A**dded, will always give the number of plots in the experiment (18 in this example).

Lead line: 3 Fertilizers × 2 Varieties (= 6 treatments) × 3 Blocks = 18 plots

Source of variation	Divisor	Degrees of freedom	Sum of squares	Mean square	Variance ratio	P
6 Treatments	18/6 = 3	5				
3 Blocks	18/3 = 6	2				
Residual		(5 × 2 =)10				
TOTAL (18 plots)		17				

Phase 1

The Correction Factor (Grand Total^2/n) is $514^2/18 = 14{,}677.6$
Added squares of the 18 numbers in the body of the table $= 15{,}964.0$
Therefore Total Sum of Squares $= 15{,}964.0 - \text{CF} = 1286.4$

Phase 2

Remember that, having calculated the total variability, we have now finished with the original data, and that Phase 2 uses only the column (Treatment) and row (Block) end totals in its calculations. These are used to calculate the sum of squares (SqADS) for Treatments and Blocks as follows below (for those who find it helpful – the full numerical calculations are given in Box 13.1). Note in SqADS that the **number** of totals to be squared and added as well as the **divisor** can now both be found respectively in the "Source of variation" and "Divisor" columns of our Anova table, and here will always multiply to the number of plots in the experiment, i.e. 18.

Sum of squares for Treatments $= \mathbf{SqA}^{6 \text{ Treatment totals}} \mathbf{D}^{\text{by } 3} \mathbf{S}^{\text{Correction Factor}}$

$$= 821.7$$

Sum of squares for Blocks $= \mathbf{SqA}^{3 \text{ Block totals}} \mathbf{D}^{\text{by } 6} \mathbf{S}^{\text{Correction Factor}} = 374.7$

BOX 13.1

Phase 2 sum of squares calculations using the Treatment (column) and Blocks (rows) totals together with the correction factor already calculated in Phase 1:

Treatment (columns) sum of squares

$$= \frac{110^2 + 116^2 + 60^2 + 74^2 + 79^2 + 75^2}{3} - 14{,}677.6 = 821.7$$

Blocks (rows) sum of squares $= \dfrac{137^2 + 173^2 + 204^2}{6} - 14{,}677.6 = 374.7$

Note that in both cases the two essential pieces of information (i.e. the number of totals to be squared and added and the divisor) can be found respectively in columns 1 and 2 of the Anova table on page 156, and that – when multiplied together – these two numbers give 18, the number of plots in the experiment.

Thus the sums of squares are partitioned as:

Lead line: 3 Fertilizers $\times$ 2 Varieties ($=$ 6 treatments) $\times$ 3 Blocks $=$ 18 plots
Correction factor: 14,677.6

Source of variation	Divisor	Degrees of freedom	Sum of squares	Mean square	Variance P ratio
6 Treatments	$18/6 = 3$	5	821.7		
3 Blocks	$18/3 = 6$	2	374.7		
Residual		$(5 \times 2 =)10$	(by subtraction) 90.0		
TOTAL (18 plots)		17	1286.4		

End Phase (of Phase 2)

At this point we need to complete the End Phase for Phase 2, since there is no point at all spending time moving on to Phase 3 if the Variance Ratio (F) for Treatments in Phase 2 does not attain significance. However, if a

Phase 3 is warranted, we will later need to complete the End Phase for the Phase 3 part of the Table (on page 160).

So (see page 99 if you need reminding) we now "go along the corridor" (to calculate Mean Squares) before "we go upstairs," using the residual mean square as the divisor to obtain Variance Ratios (F):

Lead line: 3 Fertilizers × 2 Varieties (= 6 treatments) × 3 Blocks = 18 plots
Correction factor: 14,677.6

Source of variation	Divisor	Degrees of freedom	Sum of squares	Mean square	Variance ratio	P
6 Treatments	18/6 = 3	5	⟶ 821.7	⟶ 164.3	18.3	<0.001
3 Blocks	18/3 = 6	2	⟶ 374.7	⟶ 187.4	20.8	<0.001
Residual		(5 × 2 =)10	(by subtraction)			
			⟶ 90.0	⟶ 9.0		
TOTAL (18 plots)		17	1286.4			

Even at $P = 0.001$, the Variance Ratio for Treatments exceeds the tabulated F of 10.48 for 5 (along the top of the table) and 10 (down the side) degrees of freedom, and clearly Phase 3 is warranted.

Phase 3

This is the new Phase (together with its own End Phase). Instead of splitting up the Total sum of squares into its components as we do in Phase 2, we split the Treatment sum of squares into **its** factorial components, Fertilizers, Varieties, and the Fertilizer × Variety interaction. As this means splitting up the Treatment sum of squares (calculated by SqADS of the six treatment totals), everything that we do in Phase 3 is based on these six totals. Just as we could throw away the 18 original plot data at the end of Phase 1, so we can now throw away the Block totals at the end of Phase 2, and are left with just the numbers below for Phase 3:

	AY	BY	CY	AZ	BZ	CZ	
Block 1							
Block 2							
Block 3							
	110_3	116_3	60_3	74_3	79_3	75_3	GT = 514_{18}

If we now make these numbers the data in the "body" of a new table of three Fertilizer columns by two Variety rows, we get all the numbers needed for the Phase 3 calculations:

		Fertilizers			
		A	B	C	
Varieties	Y	110_3	116_3	60_3	286_9
	Z	74_3	79_3	75_3	228_9
		184_6	195_6	135_6	GT $= 514_{18}$

This is where it really does pay to keep track of the number of data contributing to a total. In the body of the table are totals of three blocks, and now the "column" and "row" totals (outside the black frame) are SUPERTOTALS of multiples of 3. Hence the respective subscripts 6 and 9.

Allocation of degrees of freedom in Phase 3, the factorial part of the analysis

So now in Phase 3 we are going to "factorialize" a total variation in the form of the Treatment sum of squares into its components parts (Fertilizers, Varieties, and their Interaction), just as we did in Phase 2 for the Total variation of the whole experiment (into Treatments, Replicates, and their Interaction that we called the "Residual"):

Lead line: 3 Fertilizers $\times$ 2 Varieties ($=6$ treatments) $\times$ 3 Blocks $= 18$ plots
Correction factor: 14,677.6

Source of variation	Divisor	Degrees of freedom	Sum of squares	Mean square	Variance ratio	*P*
6 Treatments	$18/6 = 3$	5	821.7			

From our lead line, we see that two factors, Fertilizer and Variety, contribute to the variation in yield of our six treatments, so that these are the "sources of variation" which now replace "Treatments" and "Blocks" of Phase 2 in our Phase 3 Anova table. The degrees of freedom and "divisors" are based on information in the leadline, including 18 as the number of plots in the experiment: The Fertilizer × Variety interaction is analagous to the "residual" (see page 152 and Box 12.2).

Lead line: 3 Fertilizers × 2 Varieties (= 6 treatments) × 3 Blocks = 18 plots
Correction factor: 14,677.6

Source of variation	Divisor	Degrees of freedom	Sum of squares	Mean square	Variance P ratio
3 Fertilizers	18/3 = 6	3 − 1 = 2	We can calculate		
2 Varieties	18/2 = 9	2 − 1 = 1	We can calculate		
Fert. × Var.		2 × 1 = 2	By subtraction		
6 Treatments	18/6 = 3	5	821.7		

Just as the sum of squares will add up to the "Treatment" sum of squares (821.7), so our d.f. in the Phase 3 table add up to the "Treatment" d.f. (5) – remind yourself from Box 12.2.

Sums of squares in Phase 3, the factorial part of the analysis

As we are factorializing the sum of squares of the six "Treatment" totals, these are the totals we use but (see earlier) but grouped into larger units I call *supertotals*. Below is repeated the table of the six treatment totals within the black frame, totalled as both columns and rows to form the supertotals we need for the Phase 3 calculations.

Fertilizers

Varieties		A	B	C	
	Y	110_3	116_3	60_3	286_9
	Z	74_3	79_3	75_3	228_9
		184_6	195_6	135_6	GT = 514_{18}

The Fertilizer line in the Anova table says "3 Fertilizers" with a divisor of 6. To calculate the sum of squares for Fertilizers we have to **Sq**uare and **A**dd 3 totals and **D**ivide by 6 before **S**ubtracting the Correction Factor. I guess (if you look at the subscripts) the three totals we need from the table above are fairly obvious as 184, 195, and 135 – being the totals for fertilizers A, B, and C respectively and each being totals of six plots.

Similarly the line in the Anover table for Varieties indicates two totals with the subscript (=divisor) of 9, and 286 and 228 are easily identified as the appropriate totals of nine plots for varieties Y and Z.

BOX 13.2

The sum of squares for Fertilizers involves **SqADS** of the three totals, each of six numbers, along the bottom of the interaction totals table on page 160

$$= \frac{184^2 + 195^2 + 135^2}{6} - 14{,}677.6 = 340.1$$

The sum of squares for Varieties involves **SqADS** of the two totals, each of nine numbers, down the side of the interaction totals table.

$$= \frac{286^2 + 228^2}{9} - 14{,}677.6 = 186.8$$

Note that in both cases the two essential pieces of information (i.e. the number of totals to be squared and added and the divisor) can be found respectively in columns 1 and 2 of the Anova table on page 160, and that – when multiplied together – these two numbers give 18, the number of plots in the experiment.

The two sums of squares for Fertilizers and Varieties are thus calculated by **SqADS** and entered in the Anova table (Box 13.2 details the calculations). These two sums of squares added together account for 526.7 of the Treatment sum of squares of 821.7, leaving a **remainder** of 294.8 for the Fertilizer $\times$ Variety interaction.

Lead line: 3 Fertilizers $\times$ 2 Varieties ($=$ 6 treatments) $\times$ 3 Blocks $=$ 18 plots
Correction factor: 14,677.6

Source of variation	Divisor	Degrees of freedom	Sum of squares	Mean square	Variance ratio	P
3 Fertilizers	$18/3 = 6$	$3 - 1 = 2$	340.1			
2 Varieties	$18/2 = 9$	$2 - 1 = 1$	186.8			
Fert. $\times$ Var.		$2 \times 1 = 2$	294.8			
6 Treatments	$18/6 = 3$	5	821.7			

Having filled in the three factorial sums of squares which make up the Treatment sum of squares, it is normal to place this table on top of the Phase 2 table. Most people in fact leave out the "Treatment" line, but I find it helpful to leave it in. However, I have put it in smaller type and italics to make it clear it has to be omitted in the addition down the Table of degrees of freedom and sums of squares to the "Total":

Lead line: 3 Fertilizers × 2 Varieties (= 6 treatments) × 3 Blocks = 18 plots
Correction factor: 14,677.6

Source of variation	Divisor	Degrees of freedom	Sum of squares	Mean square	Variance ratio	P
3 Fertilizers	18/3 = 6	3 − 1 = 2	340.1			
2 Varieties	18/2 = 9	2 − 1 = 1	186.8			
Fert. × Var.		2 × 1 = 2	(by subtraction) 294.8			
6 Treatments	*18/6 = 3*	5	*821.7*	164.3	18.3	*<0.001*
3 Blocks	18/3 = 6	2	374.7	187.4	20.8	<0.001
Residual		(5 × 2 =) 10	(by subtraction) 90.0	9.0		
TOTAL (18 plots)		17	1286.4			

End Phase (of Phase 3)

As for Phase 2, we divide each sum of squares in the factorial part of the table by the appropriate degrees of freedom. The important thing to remember relates to how we calculate the variance ratios (F). Although the Fertilizer × Variety interaction was calculated like the residual in Phase 2 (as a **remainder**), it is part of the systematically designed experiment and

is NOT a residual term for calculating mean squares. Never forget: The Residual mean square from Phase 2 remains the appropriate divisor for ALL the other mean squares, whether calculated in Phase 2 or Phase 3.

Lead line: 3 Fertilizers × 2 Varieties (= 6 treatments) × 3 Blocks = 18 plots
Correction factor: 14,677.6

Source of variation	Divisor	Degrees of freedom	Sum of squares	Mean square	Variance ratio	P
3 Fertilizers	18/3 = 6	3 − 1 = 2	340.1	340.1	18.9	<0.001
2 Varieties	18/2 = 9	2 − 1 = 1	186.8	186.8	20.8	<0.001
Fert. × Var.		2 × 1 = 2	(by subtraction) 294.8	147.4	16.4	<0.001
6 Treatments	*18/6 = 3*	5	*821.7*	164.3	18.3	*<0.001*
3 Blocks	18/3 = 6	2	374.7	187.4	20.8	<0.001
Residual		(5 × 2 =) 10	(by subtraction) 90.0	9.0		
TOTAL (18 plots)		17	1286.4			

Although the effects of both Fertilizers and Varieties are highly significant at less than the 1 in 1000 chance of being an artifict of sampling, the interaction between them is similarly highly significant. Of course we still need the further tests (described on page 230 in Chapter 16) on the six "treatment" means to interpret this experiment fully, but inspection of the table of interaction totals suggests that Variety Y only outyields Z if fertilizers A or B are used, and not with fertilizer C.

Two important things to remember about factorials before tackling the next chapter

1 In my form of Anova table (unlike that in most books or in the printout from most computer programs), the "Treatment" sum of squares actually appears twice. It appears once as itself in Phase 2 and again in Phase 3 split up into three parts – the two factors and their interaction. This is useful when doing Anovas "manually," as there is no point in bothering with Phase 3 if the variance ratio (F) for "Treatments" in Phase 2 fails to attain significance.

2 The **residual mean square** is the divisor for calculating variance ratios for both Phase 2 and Phase 3 mean squares. Many beginners fall into the trap of identifying the Phase 3 interaction as a "sort of residual" since it is found by subtraction, and use its mean square as the divisor for the other Phase 3 sources of variation. You will, however, note that in my Fertilizer × Variety example in this chapter, the interaction is highly significant and clearly a result of considerable interest – indeed, it has to dominate our biological interpretation of the experiment. It is certainly not a measure of residual variation!

Analysis of factorial experiments with unequal replication

One would rarely choose to set up an experiment with this complication, but the situation can often arise that some replicates fail to produce data. I have previously raised this issue (page 116) describing the fully randomized design for analysis of variance, giving the tongue-in-cheek example of an elephant flattening one of the plots in a field experiment. I have not seen the problem of unequal replication in factorial experiments tackled in other textbooks, but I have often seen it solved in practice by the dubious procedure of discarding data until all treatments have the same number of replicates.

On page 120 I showed how the problem could be solved for completely randomized designs by abandoning the single divisor for the squared and added totals in the **SqADS** expression, and instead dividing each squared total by its own number of replicates before **A**-dding. Only one treatment had fewer replicates in that example, so I expressed the sum of squares calculation on page 121 as:

$$\frac{135^2}{3} + \frac{160^2 + 123^2}{4} - \frac{398^2}{11} = 190.22$$

Of course, this could equally be written with a separate divisor for each treatment as:

$$\frac{135^2}{3} + \frac{160^2}{4} + \frac{123^2}{4} - \frac{398^2}{11} = 190.22$$

So now let's apply this to a factorial experiment as far as necessary to illustrate the principle. The data I'll use have kindly been supplied by C. Dawson, and concern the number of offspring produced per female bean weevil from six different locations (A–F) and reared either in uncrowded (u) or crowded (c) conditions on cowpeas. Unequal replication (15 replicates were set up) resulted because of the failure of any offspring to appear in some replicates, particularly under crowded conditions. Expressing the treatment totals in an interaction table as I did earlier in this chapter for the Fertilizer × Variety experiment (page 159) is particularly helpful for the current exercise, since the little subscripts keep excellent track of variation in the number of replicates as treatment totals and then column and row totals are combined into supertotals.

		Location						
		A	B	C	D	E	F	
Rearing condition	u	982.1_{12}	1121.7_{14}	637.9_{12}	1283.1_{15}	560_{12}	1081.9_{15}	5702.7_{80}
	c	363.3_{10}	501.1_{13}	309.7_{10}	328.9_{10}	297.9_{12}	513.6_{12}	2314.5_{67}
		1345.4_{22}	1622.8_{27}	983.6_{22}	1612.0_{25}	857.9_{24}	1595.5_{27}	GT = $\mathbf{8017.2_{147}}$

This table gives us all the numbers we need for Phases 2 and 3 of the factorial analysis of variance. The penalty of using all the data, of course, is that we have to regard the treatments as "fully randomized" even if they are not, and then forego the ability of blocked designs that block variation can be separated from the residual variation.

Phase 1 is unaffected. **Total** sum of squares = all data squared and added, minus the correction factor ($8017.2^2/147$).

Phase 2 – the **treatment** sum of squares comes from the numbers within the thick black frame, allowing for the unequal replication with individual divisors as:

$$\frac{982.1^2}{12} + \frac{1121.7^2}{14} + \frac{637.9^2}{12} + \cdots + \frac{513.6^2}{12} - \text{Correction factor}$$

Phase 3 – this treatment sum of squares to 11 d.f. (for 12 treatments) can then be factorialized from the end totals of the table as:

$$\textbf{Locations} \ (5 \text{ d.f.}): = \frac{1345.4^2}{22} + \frac{1622.8^2}{27} + \frac{983.6^2}{12} + \cdots$$

$$+ \frac{1595.5^2}{27} - \text{Correction factor}$$

$$\textbf{Rearing condition} \ (1 \text{ d.f.}): = \frac{5702.7^2}{80} + \frac{2314.5^2}{67} - \text{Correction factor}$$

Interaction (5 d.f.): **Treatment** sum of squares *minus* **Location** sum of squares *minus* **Rearing condition** sum of squares.

In Chapter 16 (page 214), I go on to explain how to judge differences between two means with different replication, but if you go back to page 68 and the *t*-test, you'll already find the answer there!

EXECUTIVE SUMMARY 6
Analysis of a 2-factor randomized block experiment

In a factorial experiment, the "treatments" are combinations of more than one experimental variable, e.g. 12 "treatments" might result from all combinations of four varieties and three spacings. This is a very common form of design, and (in the example just given) interest centers not only on how the varieties perform and the overall effect of spacing, but also on the **INTERACTION** of varieties × spacing – i.e. do all the varieties react to the different spacings in the same way and to the same degree? If not, then we have **INTERACTION**.

Phases 1 and 2 (see Executive Summary 5, page 139)

PHASE 1 is **the phase of original data**. We first use the original data (i.e. the figures in the body of the table) to obtain the **total sum of square**$_{\text{of deviations}}$ (=added squares − correction factor).

PHASE 2 is **the phase of treatment and replicate totals**. We use the end totals of rows and columns – first the treatment, then the replicate totals (remember **SqADS**) to calculate the **treatment** and **replicate** sums of . . . squares$_{\text{of deviations}}$. The **residual** sum of . . . squares$_{\text{of deviations}}$ is then found by subtracting these two sums of squares from the total sum of . . . squares$_{\text{of deviations}}$.

We then complete this part of the analysis with the "end phase": IT IS NOT WORTH GOING ON TO PHASE 3 UNLESS THE "*F*" VALUE FOR TREATMENTS IS SIGNIFICANT.

Phase 3 (a new phase!) – The phase of treatment "supertotals"

This phase involves subdividing the "treatment" sum of squares$_{\text{of deviations}}$ into its components. All the work in this phase uses just the TREATMENT TOTALS, NOT INDIVIDUALLY – BUT ADDED TOGETHER IN VARIOUS COMBINATIONS into **SUPERTOTALS**. In the simple 2-variable experiment given as an example above, these components (imagining 3 replicates of the 12 treatments and therefore 36 plots) would be as follows (with the thinking involved and the calculation procedure).

It is useful to set out the **treatment totals** as a grid of one factor against the other, the column and row totals will now give us the **supertotals** we need for SqADS. If we also put in the number of plots involved in each total as subscripts, beginning with 3 (=the number of replicates) for the treatment totals used in Phase 2, then we can also sum these and keep our eye on the correct divisor in SqADS!

	V1	V2	V3	V4	
Spacing 1	Tot_3	Tot_3	Tot_3	Tot_3	$Supertot_{12}$
Spacing 2	Tot_3	Tot_3	Tot_3	Tot_3	$Supertot_{12}$
Spacing 3	Tot_3	Tot_3	Tot_3	Tot_3	$Supertot_{12}$
	$Supertot_9$	$Supertot_9$	$Supertot_9$	$Supertot_9$	

4 varieties (4 supertotals to SqA, D by 9*, S the correction factor. Degrees of freedom will be number varieties $- 1 = 3$).

*D by 9: Each variety supertotal is the total of nine original plot figures – see subscripts (with three replicates this means the supertotal of three "treatment" totals. Check: 4 supertotals $\times 9 = 36$.)

CALCULATION: 4 supertotals to SqA $-$ D(by 9) $-$ S(correction factor), i.e.:

$$\frac{()^2 + ()^2 + ()^2 + ()^2}{9} - \text{Correction factor}$$

where each set of brackets contains the supertotal for a different variety, i.e. the sum of the three treatment totals which involve that particular variety.

3 spacings (3 supertotals to SqA, D by 12*, S the correction factor. Degrees of freedom will be number spacings $- 1 = 2$).

*D by 12: Each spacing supertotal is the total of 12 original plot figures – see subscripts (with three replicates this means the supertotal of four "treatment" totals. Check: 3 supertotals $\times 12 = 36$).

CALCULATION: 3 supertotals to SqA $-$ D(by 12) $-$ S(correction factor), i.e.:

$$\frac{()^2 + ()^2 + ()^2}{12} - \text{Correction factor}$$

where each set of brackets contains the supertotal for a different spacing, i.e. the sum of the four treatment totals which involve that particular spacing.

4 variety $\times$ 3 spacings interaction: Sum of squares *of deviations* obtained by subtracting variety and spacing sums of squares *of deviations* from the **TREATMENT** sum of squares *of deviations* (NB: INTERACTIONS

ARE ALWAYS "**REMAINDERS**"). Degrees of freedom are (d.f. for varieties) × (d.f. for spacing) = 3 × 2 = 6: check: Degrees of freedom for varieties, spacings, and interaction add up to treatment degrees of freedom.

End Phase

1 Calculate variance (=**mean square**) by working **HORIZONTALLY** across the analysis of variance table to divide each sum of squares by its own degrees of freedom.
2 Calculate **variance ratio** (*F*) by working **VERTICALLY** upwards in the mean square column, dividing ALL mean squares by the "residual mean square" calculated in PHASE 2.
3 Check the significance of the *F* values with statistical tables.

Spare-time activity

The yields of four white grape varieties (Müller–Thurgau, Semillon, Pinot Blanc, and Huchselrebe) were compared, using traditional training and the high-wire system for each variety. There were five replicates (I–V) of the experiment. The yields (tonnes/ha) were as follows:

		Müller-Th.	Semillon	Pinot B.	Huchselr.
I	Traditional	4.73	4.65	8.70	10.78
	High-wire	3.36	6.02	9.29	13.13
II	Traditional	3.23	6.29	8.69	6.95
	High-wire	2.10	5.92	8.42	10.09
III	Traditional	4.69	6.12	7.41	8.91
	High-wire	6.99	8.33	8.52	11.21
IV	Traditional	5.21	6.48	9.64	9.04
	High-wire	4.78	9.52	10.11	11.53
V	Traditional	4.92	7.42	10.16	6.16
	High-wire	3.21	9.01	10.61	9.06

Carry out an analysis of variance. Is there statistical validity in the idea the table suggests that different varieties do better on each of the two training systems?

Keep your calculations. A further exercise in the interpretation of this experiment will be found in the "spare-time activities" following Chapter 16, which explains how differences between means should be evaluated in a factorial analysis.

14

Factorial experiments with more than two factors – leave this out if you wish!

Chapter features

Introduction

The previous chapter developed the analysis of a factorial experiment with just two factors, using an example of fertilizers and varieties. With two factors, only one interaction is possible, and we were able to calculate this as a remainder from the overall "Treatment" sum of squares in Phase 3. This introduced the important principle that we calculate the sum of squares for interactions as a **remainder** from a larger sum of squares which includes it as well as the sums of squares of the sources of variation that are interacting. However, Chapter 13 did not equip us to handle the calculation of interaction sums of squares in more complex factorial experiments which may include several or even many interactions.

We will jump straight to an example of a factorial experiment with four factors. If you can handle that, you can probably handle any factorial! You may wish to stop reading now and go straight to Chapter 15 – for practical reasons very few people ever do 4-factor experiments. It might even be unwise, seeing that my example has 36 treatments per replicate. This (see Chapter 11) is asking for trouble in terms of residual variation across the experiment, and one of the incomplete designs or a split-plot design (see next chapter) might well be preferable.

However, it is worth understanding how to calculate interaction sums of squares (still always as **remainders** from a larger figure) in such experiments. To stop things becoming too tedious for you, I will try to cope with the principles with an example or two of the calculation procedures involved, and relegate the rest of the calculations to the optional addendum at the end of the chapter.

Different "orders" of interaction

The main factors, e.g. A, B, C, are called just that – *main factors.*

Interactions of two factors, e.g. A × B or A × D, are called – *first order interactions.*

Interactions of three factors, e.g. A × B × C or B × C × D, are called – *second order interactions.*

Interactions of four factors, e.g. A × B × C × D or B × C × D × E, are called – *third order interactions.*

Do I need to go on?

It is quite useful to be able to work out how many interactions of each "order" will feature in a complex experiment. Then we can be sure we have identified them all. This topic is covered in Box 14.1.

It is only the sum of squares for the highest order of interaction possible in an experiment that is finally obtained as a remainder from the overall "Treatment" sum of squares; the other interaction sums of squares are still calculated as remainders, and the skill of analyzing a complex

---

BOX 14.1   The interactions involved in factorial experiments

Working this out involves the formula for "combinations," a permutation to be found in many statistical or mathematical textbooks.

First you need to understand what the mathematical (not the statistical) term "factorial" of a number means. In mathematics, it is the product of a descending series of numbers – e.g. factorial 7 is 7 × 6 × 5 × 4 × 3 × 2 × 1 = 5040. The standard notation is an exclamation mark, e.g. factorial 7 is represented by **7!**

Secondly, we need to distinguish between the total number of factors in the entire experiment (I'll term this **"total factors"**) and the number of factors in any order of interaction (I'll term this **"set size"** – e.g. this would be three for second order interactions such as A × B × C).

With these two definitions, the number of possible combinations for any order of interaction (set size) is:

$$\frac{\text{“factorial” of } \textbf{total factors}}{\text{“factorial” of } \textbf{set size} \times \text{“factorial” of (\textbf{total factors – set size})}}$$

(Continued)

BOX 14.1 Continued

This formula is provided in case you should ever need it! To make that extremely unlikely, I have worked out for you all the combinations for up to eight factors, and I don't think I've ever encountered an experiment with that many!

Number of factors	Order of interaction						
	1st	2nd	3rd	4th	5th	6th	7th
2	1						
3	3	1					
4	6	4	1				
5	10	10	5	1			
6	15	20	15	6	1		
7	21	35	35	21	7	1	
8	28	56	70	56	28	8	1

factorial experiment is to work out of which sum of squares they are the remainder!

Example of a 4-factor experiment

The experiment which forms the example for this chapter is about leeks. Varieties and fertilizers are again involved, but the added factors are planting leeks with the leaf fan along the row or at right angles and trimming or not trimming the roots and leaves before planting. These last two factors are testing gardening "folk lore." The data are marketable yield (kg per plot) and the factors and their levels (with code to be used subsequently) are as follows:

Varieties (3 levels) : A, B and C

Fertilizer (2 levels): A high nitrogen fertilizer applied as a top dressing once (1) or twice as a split dose (2)

Planting alignment (2 levels): Parallel (P) or at right angles (Q) to the row. *Folk lore suggests that leek leaf fans facing the sun (alignment P in this trial) will produce higher yields than leeks with their fans at right angles to the sun (Q).*

Trimming (3 levels): Roots (R) or leaves (L) or both roots and leaves (T) trimmed. *Folk lore suggests that trimming both the tops and roots of leeks at planting raises yields.*

The experiment was planted as a $3 \times 2 \times 2 \times 3$ factorial (i.e. 36 plots) at three sites (representing randomized blocks), making 108 plots in all.

Prelims

The data from the randomization are not presented; I have already unscrambled it in preparation for Phases 1 and 2, with subscripts to show the number of plots contributing to each total. **But take special notice** – because of the large number of treatments, for the first time in the book (in order to fit the page format) the treatments this time are the rows in the table below, with the few blocks the columns. Previously the tables of data have been presented with the treatments as the columns (e.g. page 155).

Var	Fert	Align	Trim	Blocks 1	2	3	Total
A	1	P	R	8.3	7.4	9.8	25.5_3
A	1	P	L	7.1	6.2	9.2	22.5_3
A	1	P	T	9.4	7.8	9.8	27.0_3
A	1	Q	R	6.9	6.8	9.1	22.8_3
A	1	Q	L	6.9	5.7	8.4	21.0_3
A	1	Q	T	7.0	5.9	8.7	21.6_3
A	2	P	R	7.9	7.5	11.6	27.0_3
A	2	P	L	7.5	6.5	9.4	23.4_3
A	2	P	T	8.9	7.6	11.1	27.6_3
A	2	Q	R	7.2	6.2	9.1	22.5_3
A	2	Q	L	6.9	6.0	8.7	21.6_3
A	2	Q	T	7.8	6.4	8.7	22.9_3
B	1	P	R	7.8	7.1	11.3	26.2_3
B	1	P	L	8.2	7.1	10.3	25.6_3
B	1	P	T	8.5	7.3	10.6	26.4_3
B	1	Q	R	5.8	5.0	7.2	18.0_3
B	1	Q	L	5.5	4.7	6.9	17.1_3
B	1	Q	T	5.4	5.2	7.4	18.0_3

(Continued)

Continued

| Var | Fert | Align | Trim | Blocks | | | Total |
				1	2	3	
B	2	P	R	9.5	7.9	10.2	27.6_3
B	2	P	L	9.2	9.2	10.1	28.5_3
B	2	P	T	9.0	8.9	10.5	28.4_3
B	2	Q	R	5.5	5.4	7.7	18.6_3
B	2	Q	L	6.0	6.0	5.9	17.9_3
B	2	Q	T	6.6	5.2	7.4	19.2_3
C	1	P	R	9.3	8.6	9.1	27.0_3
C	1	P	L	8.6	7.2	10.0	25.8_3
C	1	P	T	9.0	8.8	9.9	27.7_3
C	1	Q	R	8.9	8.4	9.6	26.9_3
C	1	Q	L	6.9	6.7	11.9	25.5_3
C	1	Q	T	9.4	8.6	9.5	27.5_3
C	2	P	R	9.9	8.5	10.4	28.8_3
C	2	P	L	8.7	7.6	10.7	27.0_3
C	2	P	T	10.0	8.4	10.3	28.7_3
C	2	Q	R	7.8	7.6	12.8	28.2_3
C	2	Q	L	9.4	9.2	8.4	27.0_3
C	2	Q	T	9.6	8.2	10.8	28.6_3
Block totals				286.3_{36}	256.8_{36}	342.5_{36}	**GT = 885.6_{108}**

Next we compose the lead line and the skeleton analysis table (at this stage just for the 36 "treatments" ignoring their factorial nature) with degrees of freedom $(n-1)$. The "divisor" for SqADS, multiplied by the number of totals to be **Sq**uared and **A**dded (written in the table below in front of the "source of variation"), always gives the number of plots in the experiment (108 in this example).

Lead line: 3 Varieties × 2 Fertilizer levels × 2 Alignments × 3 Trimmings (= 36 treatments) × 3 Blocks = 108 plots

Source of variation	Divisor	Degrees of freedom	Sum of squares	Mean square	Variance ratio	P
36 Treatments	$108/36 = 3$	35				
3 Blocks	$108/3 = 36$	2				
Residual		$(35 \times 2 =)$ 70				
TOTAL (108 plots)		107				

Phase 1

The Correction Factor (Grand Total$^2/n$) is $885.6^2/108 = 7261.92$
All the 108 numbers in the body of the table squared and added together
$= 7565.84$
Therefore Total Sum of Squares $= 7565.84 - CF = 303.92$

Phase 2

By now, you will probably not need a further reminder about SqADS for the
36 Treatments and 3 Blocks, so I will move straight on to the Phase 2 analysis of variance table, complete with the End Phase for Phase 2 (i.e. mean
squares and variance ratios). If I am being optimistic, then Box 14.2 covers
the SqADS bit.

Lead line: 3 Varieties × 2 Fertilizer levels × 2 Alignments × 3 Trimmings (= 36 treatments) × 3
Blocks = 108 plots
Correction factor = 7261.92

Source of variation	Divisor	Degrees of freedom	Sum of squares	Mean square	Variance ratio	P
36 Treatments	108/36 = 3	35	→ 157.93	→ 4.51	7.70	<0.001
3 Blocks	108/3 = 36	2	→ 105.31	→ 52.65	90.63	<0.001
Residual		(35 × 2 =) 70	→ 40.68	→ 0.58		
TOTAL (108 plots)		107	303.92			

BOX 14.2

Phase 2 uses only the row (Treatment in this example) and column (Blocks)
end totals. These are used to calculate the sum of squares (SqADS) for
Treatments and Blocks as follows:

Sum of squares for Treatments = **SqA**$^{36 \text{ Treatment totals}}$**D**$^{\text{by } 3}$**S**$^{\text{Correction Factor}}$

$$= \frac{25.5^2 + 22.5^2 + 27.0^2 + 22.8^2 + 21.0^2 + \cdots + 28.6^2}{3} - 7261.92 = 157.93$$

Sum of squares for Blocks = **SqA**$^{3 \text{ Block totals}}$**D**$^{\text{by } 36}$**S**$^{\text{Correction Factor}}$

$$= \frac{286.3^2 + 256.8^2 + 342.5^2}{36} - 7261.92 = 105.31$$

Phase 3

We now come to the meat of this chapter – dealing with the manifold interactions between the main factors – Varieties (V), Fertilizers (F), Alignments (A), and Trimmings (T) – in this complex factorial experiment.

The sources of variation in Phase 3

Now we list the factorial sources of variation in preparation for the factorial part of the analysis of variance table. Each line begins with the number of the supertotals we will need to identify.

Main effects.

Number of "levels" of the factor	Source of variation	Divisor	Degrees of freedom
	3 Varieties (V)	108/3 = 36	2
	2 Fertilizers (F)	108/2 = 54	1
	2 Alignments (A)	108/2 = 54	1
	3 Trimmings (T)	108/3 = 36	2

First order interactions. Each of the four main effects above can interact with each of the three others (Variety × Fertilizer, Fertilizer × Trimmings, etc.). The number of supertotals we will need for each interaction table is found by multiplying the number for the two factors (we'll just use the letter code from the table immediately above) involved in the interaction. Reminder – the d.f. are similarly found by multiplying the degrees of freedom for the factors interacting. Again the divisor (for SqADS) is obtained by dividing the total number of plots in the experiment (108) by that number of totals, and that divisor (remember?) is also the number of plots contributing to each supertotal.

The table below gives all the possible pairings of three factors:

Number of levels	Source of variation	Divisor	Degrees of freedom
	3 × 2 V × F	108/6 = 18	$(3-1) \times (2-1) = 2$
	3 × 2 V × A	108/6 = 18	$(3-1) \times (2-1) = 2$
	3 × 3 V × T	108/9 = 12	$(3-1) \times (3-1) = 4$
	2 × 2 F × A	108/4 = 27	$(2-1) \times (2-1) = 1$
	2 × 3 F × T	108/6 = 18	$(2-1) \times (3-1) = 2$
	2 × 3 A × T	108/6 = 18	$(2-1) \times (3-1) = 2$

Note the systematic way in which the interactions are listed, to make sure we miss none! We begin with all interactions with Variety, then with

Fertilizer (other than F × V, which would repeat V × F) and finally the only nonrepeat with Alignment is A × T.

Second order interactions.   These are interactions between three main factors, and the process for working out the number of supertotals, the divisor, and the d.f. remains the same. With four factors (see Box 14.1) there are just four different combinations of any three factors:

	Source of variation	Divisor	Degrees of freedom
Number of levels →	3 × 2 × 2 V × F × A	108/12 = 9	$(3-1) \times (2-1) \times (2-1) = 2$
	3 × 2 × 3 V × F × T	108/18 = 6	$(3-1) \times (2-1) \times (3-1) = 4$
	3 × 2 × 3 V × A × T	108/18 = 6	$(3-1) \times (2-1) \times (3-1) = 4$
	2 × 2 × 3 F × A × T	108/12 = 9	$(2-1) \times (2-1) \times (3-1) = 2$

Third order interaction.   There is only one third order interaction possible between four factors $-$V × F × A × T. This involves $3 \times 2 \times 2 \times 3 = 36$ supertotals, each of $108/36 = 3$ plots and with $(3-1) \times (2-1) \times (2-1) \times (3-1) = 4$ d.f. Does 36 totals of three numbers ring a bell? It should do. These are the 36 row treatment totals (each the sum of three replicates) we have alresady used for SqADS to calculate the "Treatment sum of squares" in Phase 2 (page 175), leaving the four factor interaction as the **ultimate** remainder!

Factorial part of the analysis table

We can therefore now construct the full factorial part of the Anova table, by putting together the main factors and all the different interactions we have just identified, together with the number of levels for each, the divisors for the **D** in **SqADS**, and the respective degrees of freedom. You will see that all the d.f. add up to 35, the d.f. for Treatments in the Phase 2 analysis table (page 175):

Lead line: 3 Varieties × 2 Fertilizer levels × 2 Alignments × 3 Trimmings (= 36 treatments) × 3 Blocks = 108 plots
Correction factor = 7261.92

Source of variation	Divisor	Degrees of freedom	Sum of squares	Mean square	Variance ratio	P
Main effects						
3 Varieties (V)	108/3 = 36	2				
2 Fertilizers (F)	108/2 = 54	1				
2 Align (A)	108/2 = 54	1				
3 Trim (T)	108/3 = 36	2				

(Continued)

Continued

Source of variation	Divisor	Degrees of freedom	Sum of squares	Mean square	Variance ratio	P
1st order interactions						
3 × 2 V × F	108/6 = 18	2				
3 × 2 V × A	108/6 = 18	2				
3 × 3 V × T	108/9 = 12	4				
2 × 2 F × A	108/4 = 27	1				
2 × 3 F × T	108/6 = 18	2				
2 × 3 A × T	108/6 = 18	2				
2nd order interactions						
3 × 2 × 2 V × F × A	108/12 = 9	2				
3 × 2 × 3 V × F × T	108/18 = 6	4				
3 × 2 × 3 V × A × T	108/18 = 6	4				
2 × 2 × 3 F × A × T	108/12 = 9	2				
3rd order interaction						
V × F × A × T		4				
36 Treatments	108/36 = 3	35	157.93	4.51	7.70	<0.001

Note: The **bottom** line here repeats the **top** line of the Phase 2 analysis table (*page 175*).

The sums of squares in the factorial part of the analysis

This job can be tackled in a variety of ways, but paradoxically I recommend not beginning at the top, but with the first order interactions. This is because the supertotals for SqADS for the main factors will be calculated as part of this process (see diagram below), so basically we are doing main effects and first order interactions at the same time! It is actually even more efficient to begin with 2nd order interactions, but in the cause of your sanity I advise against it!

Two important reminders:

1 Phase 3 uses only the 36 treatment totals (the Row totals here), combined into various **supertotals** before SqADS.
2 The *supertotals* we will need to calculate sums of squares for first order interactions are set up as interaction tables (e.g. page 180). The body of the table provides the total sum of squares of these interaction totals, while the row and column totals provide component sums of squares to be deducted from the total to leave the first order inter-action sum of squares as a **remainder**. This is illustrated graphically

below (cf. also page 152); for higher order interactions I fear still further steps are necessary (see later).

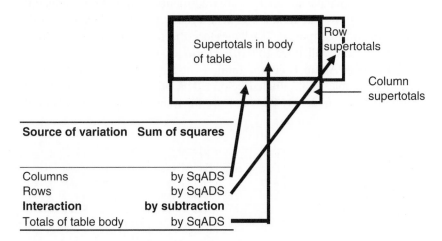

Source of variation	Sum of squares
Columns	by SqADS
Rows	by SqADS
Interaction	**by subtraction**
Totals of table body	by SqADS

It's probably a good idea to tabulate the treatment totals from the table of data under "Prelims" (page 173) in a more convenient way than 36 row end totals. The table below is so arranged that all matching levels of Variety (A, B, or C) and Fertilizer (1 or 2) are in the same vertical column (e.g. all A1s on the far left) and all matching levels of Alignments (P or Q) and Trimmings (R, L or T) are along the same row (e.g. PR in the top row):

A1PR	25.5_3	A2PR	27.0_3	B1PR	26.2_3	B2PR	27.6_3	C1PR	27.0_3	C2PR	28.8_3
A1PL	22.5_3	A2PL	23.4_3	B1PL	25.6_3	B2PL	28.5_3	C1PL	25.8_3	C2PL	27.0_3
A1PT	27.0_3	A2PT	27.6_3	B1PT	26.4_3	B2PT	28.4_3	C1PT	27.7_3	C2PT	28.7_3
A1Q	22.8_3	A2Q	22.5_3	B1QR	18.0_3	B2QR	18.6_3	C1QR	26.9_3	C2QR	28.2_3
A1QL	21.0_3	A2QL	21.6_3	B1QL	17.1_3	B2QL	17.9_3	C1QL	25.5_3	C2QL	27.0_3
A1QT	21.6_3	A2QT	22.9_3	B1QT	18.0_3	B2QT	19.2_3	C1QT	27.5_3	C2QT	28.6_3

Sum of squares for first order interactions and main effects

Let's therefore begin with the interaction *Varieties (V) × Fertilizers (F)*. Our Anova table tells us this is an interaction of 3×2 levels, meaning we set up our interaction table as 3 columns (varieties) × 2 rows (fertilizers).

	Variety A	Variety B	Variety C	
Fertilizer 1				
Fertilizer 2				
				885.6_{108}

With six "boxes" in the body of the table and 108 plots in total, each box will contain a total of $108/6 = 18$ plots (each total will be the **supertotal** of the three replicates for each of six treatments) and will have added the subscript 18 (see next table). You will already find this divisor identified in the Anova table (page 178). Therefore the total for box A × 1 in our interaction table will have the **supertotal** of the six treatments which share the codes A and 1, i.e. the total of the left hand column in our new "treatment" table total (shaded below). The other five supertotals of six treatment totals (A2, B1, B2, etc.) are fairly obviously the totals of the other five vertical columns:

A1PR	25.5_3	A2PR	27.0_3	B1PR	26.2_3	B2PR	27.6_3	C1PR	27.0_3	C2PR	28.8_3
A1PL	22.5_3	A2PL	23.4_3	B1PL	25.6_3	B2PL	28.5_3	C1PL	25.8_3	C2PL	27.0_3
A1PT	27.0_3	A2PT	27.6_3	B1PT	26.4_3	B2PT	28.4_3	C1PT	27.7_3	C2PT	28.7_3
A1Q	22.8_3	A2Q	22.5_3	B1QR	18.0_3	B2QR	18.6_3	C1QR	26.9_3	C2QR	28.2_3
A1QL	21.0_3	A2QL	21.6_3	B1QL	17.1_3	B2QL	17.9_3	C1QL	25.5_3	C2QL	27.0_3
A1QT	21.6_3	A2QT	22.9_3	B1QT	18.0_3	B2QT	19.2_3	C1QT	27.5_3	C2QT	28.6_3

So our interaction table becomes:

	Variety A	Variety B	Variety C	
Fertilizer 1	140.4_{18}	131.3_{18}	160.4_{18}	432.1_{54}
Fertilizer 2	145.0_{18}	140.2_{18}	168.3_{18}	453.5_{54}
	285.4_{36}	271.5_{36}	328.7_{36}	$\mathbf{885.6_{108}}$

We can now **SqA**the appropriate supertotals in the table **D**by the subscript value **S**the Correction Factor the six **supertotals**(each of 18 plots) in the body of the table to obtain 53.96 as the sum of squares for the table – a *combination* of the sums of squares for Varieties *plus* Fertilizers *plus* their Interaction (the interaction component will finally be found as a remainder by subtraction).

Now we see the advantage of starting with a 2-factor interaction. In the end totals of this little table we already have (in our column and row totals respectively) the three **supertotals** of 36 plots (285.4, 271.5, and 328.7) that our Anovar table tells us that we need to SqADS for the sum of squares for the Varieties main effect, and also the two **supertotals** of 54 plots (432.1 and 455.3) that we need for the sums of squares for the

Fertilizers main effect:

Correction factor $= 7261.92$

Source of variation	Sum of squares
Columns	by SqADS
Rows	by SqADS
Interaction	**by subtraction**
Totals of table body	by SqADS

Source of variation	Sum of squares
Varieties	49.44
Fertilizer	4.24
Interaction	**0.28**
Totals of table body	53.96

To save being repetitive, the working for the sums of squares for most of the remaining first order interactions is given in the Addendum at the end of this chapter, but it is probably worth elaborating a second example of first order interaction (F $\times$ A) below, but condensed into note form based on our first example.

Interaction table: 2 columns (Fertilizer levels) $\times$ 2 rows (Alignment levels) $= 4$ *supertotals*

Divisor (number of plots per supertotal in body of table): $108/4 = 27$

Number of treatment totals$_3$ per supertotal in body of table: $27/3 = 9$

Identification of first such supertotal: Pick one Fertilizer and one Alignment, say Fertilizer 1 and Alignment P. Treatments sharing 1 and P are dark shaded in table below. If we also shade (lighter) the totals sharing 2 and Q, it is clear how the 36 treatment totals divide into the four supertotals of nine treatment totals we need. Each *supertotal* involves summing the totals in three blocks of three totals in the table below:

A1PR	**25.5_3**	A2PR	27.0_3	**B1PR**	**26.2_3**	B2PR	27.6_3	**C1PR**	**27.0_3**	C2PR	28.8_3
A1PL	**22.5_3**	A2PL	23.4_3	**B1PL**	**25.6_3**	B2PL	28.5_3	**C1PL**	**25.8_3**	C2PL	27.0_3
A1PT	**27.0_3**	A2PT	27.6_3	**B1PT**	**26.4_3**	B2PT	28.4_3	**C1PT**	**27.7_3**	C2PT	28.7_3
A1QR	22.8_3	**A2QR**	**22.5_3**	B1QR	18.0_3	**B2QR**	**18.6_3**	C1QR	26.9_3	**C2QR**	**28.2_3**
A1QL	21.0_3	**A2QL**	**21.6_3**	B1QL	17.1_3	**B2QL**	**17.9_3**	C1QL	25.5_3	**C2QL**	**27.0_3**
A1QT	21.6_3	**A2QT**	**22.9_3**	B1QT	18.0_3	**B2QT**	**19.2_3**	C1QT	27.5_3	**C2QT**	**28.6_3**

Supertotals in interaction table:

	Fertilizer 1	Fertilizer 2	
Alignment P	233.7_{27}	247.0_{27}	480.7_{54}
Alignment Q	198.4_{27}	206.5_{27}	404.9_{54}
	432.1_{54}	453.5_{54}	**885.6_{108}**

Table of sum of squares (SqADS):

Correction factor $= 7261.92$

Source of variation	Sum of squares		Source of variation	Sum of squares
Columns (Fertilizer)	Already calculated for earlier table	➡	Fertilizer	4.24
Rows (Alignment)	by SqADS	➡	Alignment	53.20
Interaction	**by subtraction**		**Interaction**	**0.25**
Totals of table body	by SqADS	➡	Totals of table body	57.69

Sums of squares for second order interactions

We will begin (taking the 3-factor interactions in order from the table on page 178) with the interaction $V \times F \times A$ between Variety (three levels), Fertilizer (two levels), and Alignment (two levels). Our interaction table now looks rather more complex, as three factors are involved:

	Variety A		Variety B		Variety C		
	Fert. 1	Fert. 2	Fert. 1	Fert. 2	Fert. 1	Fert. 2	
Alignment P							
Alignment Q							
							885.6_{108}

Inside the black frame are the spaces for the $3 \times 2 \times 2 = 12$ supertotals we will need to SqADS. The 12 totals from 108 data and the analysis table on page 178 remind us the divisor is $108/12 = 9$ (i.e. each supertotal is made up of $9/3 = 3$ treatment totals, since each treatment total is already a total of 3 plots). We start by identifying the three treatment totals which make up the 12 supertotals we are looking for by dark shading the treatments which share A, 1, and P in the table below. If we also light shade those that share A, 2, and Q, it is clear that each column of the table below provides two supertotals.

A1PR	**25.5₃**	A2PR	27.0_3	B1PR	26.2_3	B2PR	27.6_3	C1PR	27.0_3	C2PR	28.8_3
A1PL	**22.5₃**	A2PL	23.4_3	B1PL	25.6_3	B2PL	28.5_3	C1PL	25.8_3	C2PL	27.0_3
A1PT	**27.0₃**	A2PT	27.6_3	B1PT	26.4_3	B2PT	28.4_3	C1PT	27.7_3	C2PT	28.7_3
A1QR	22.8_3	**A2QR**	**22.5₃**	B1QR	18.0_3	B2QR	18.6_3	C1QR	26.9_3	C2QR	28.2_3
A1QL	21.0_3	**A2QL**	**21.6₃**	B1QL	17.1_3	B2QL	17.9_3	C1QL	25.5_3	C2QL	27.0_3
A1QT	21.6_3	**A2QT**	**22.9₃**	B1QT	18.0_3	B2QT	19.2_3	C1QT	27.5_3	C2QT	28.6_3

We can now do the addition and complete the table of supertotals:

	Variety A		Variety B		Variety C		
	Fert. 1	Fert. 2	Fert. 1	Fert. 2	Fert. 1	Fert. 2	
Alignment P	75.0_9	78.0_9	78.2_9	84.5_9	80.5_9	84.5_9	480.7_{54}
Alignment Q	65.4_9	67.0_9	53.1_9	55.7_9	79.9_9	83.8_9	404.9_{54}
	140.4_{18}	145.0_{18}	131.3_{18}	140.2_{18}	160.4_{18}	168.3_{18}	885.6_{108}

If we SqADS the supertotals in the body of the table, we get a sum of squares of **146.94**. This is a large proportion of the entire Treatment sum of squares of 157.93 we calculated in Phase 2, and of course only a small **remainder** of the 146.94 is attributable to the interaction between all three factors. Most of the variation of the supertotals in the body of the table comes from variation we have already calculated, the main effects of the three factors involved, and that of all first order interactions (between any two of the three factors). So, unfortunately we have to abandon our easy way of getting at the interaction remainder (based on row and column totals of the above interaction table), and instead list all the components contributing to our sum of squares of 146.94, together with their sum of squares as already calculated:

Source of variation	Sum of squares
Variety	49.44
Fertilizer	4.24
Alignment	53.20
V × F (remainder)	0.28
F × A (remainder)	0.25
V × A (remainder)	39.34
Interaction	**by subtraction** $= 0.19$
Totals of table body	146.94

Subtracting all the main factor and first order interaction components from 146.94 leaves a small **remainder** (as predicted above) of **0.19** for the V × F × A interaction.

To avoid tedious repetition for those who don't need it, the workings for the other three second order interactions is relegated to the addendum to this chapter.

To the End Phase

Once we have completed all the calculations, we can:

- Combine the Phase 2 and Phase 3 Anova tables and fill in all the calculated sums of squares.
- Find the missing sum of squares for the third order interaction of all four factors by subtracting all the Main Effect and Interaction sum of squares calculated in Phase 3 from the "Treatment sum of squares" from Phase 2. Note that I have again put the latter in small type to avoid the impression of "double accounting" since its sums of squares and its degrees of freedom re-appear as the main effect and interactions above it.

- Divide each sum of squares by its degrees of freedom to obtain the mean squares
- Find the Variance Ratios (F) by dividing all mean squares by the **Residual mean square (= 0.58) in the Phase 2 part of the analysis table**. The Residual mean square from Phase 2 is used throughout; even the third order interaction is a component of the experimental treatments and might be important.
- Check the significance of the size of the variance ratios in F tables with 70 as the residual d.f.

So here is the completed Anova table for this 4-factor experiment (ns = not significant at $P = 0.05$).

Lead line: 3 Varieties × 2 Fertilizer levels × 2 Alignments × 3 Trimmings (= 36 treatments) × 3 Blocks = 108 plots
Correction factor = 7261.92

Source of variation	Divisor	Degrees of freedom	Sum of Squares	Mean square	Variance ratio	P
Main effects						
3 Varieties (V)	108/3 = 36	2	49.44	24.72	42.62	<0.001
2 Fertilizers (F)	108/2 = 54	1	4.24	4.24	7.31	<0.01
2 Alignments (A)	108/2 = 54	1	53.20	53.20	91.72	<0.001
3 Trimmings (T)	108/3 = 36	2	6.59	3.30	5.69	<0.01
1st order interactions						
3×2 V×F	108/6 = 18	2	0.28	0.14	0.93	ns
3×2 V×A	108/6 = 18	2	39.34	19.67	33.91	<0.001
3×3 V×T	108/9 = 12	4	1.72	0.43	0.74	ns
2×2 F×A	108/4 = 27	1	0.25	0.25	0.43	ns
2×3 F×T	108/6 = 18	2	0.03	0.02	0.03	ns
2×3 A×T	108/6 = 18	2	0.39	0.20	0.34	ns
2nd order interactions						ns
3×2×2 V×F×A	108/12 = 9	2	0.19	0.10	0.17	ns
3×2×3 V×F×T	108/18 = 6	4	0.16	0.04	0.07	ns
3×2×3 V×A×T	108/18 = 6	4	1.72	0.43	0.74	ns
2×2×3 F×A×T	108/12 = 9	2	0.14	0.07	0.12	ns
3rd order interaction			by subtraction			
V×F×A×T		4	0.24	0.06	0.10	ns
36 Treatments	108/36 = 3	35	157.93	4.51	7.70	<0.001
3 Blocks	108/3 = 36	2	105.31	52.65	90.63	<0.001
Residual		(35 × 2 =) 70	40.68	0.58	Used as divisor for all variance ratios	
TOTAL (108 plots)		107	303.92			

We still need to make further tests (described in Chapter 16) in order to identify exactly how the yield of our leeks has been affected by our factorial treatment combinations; this chapter – like the others before it concerning analysis of variance – restricts itself to the analysis of variance procedure.

However, it is perhaps worthwhile just to look at the likely interpretation of the interaction table for the one significant first order interaction,

Varieties × Alignment. This time, the supertotals have been converted to their mean values as kg/plot, with a superscript showing how many plots have been averaged:

	Variety A	Variety B	Variety C	
Alignment P	8.50^{18}	9.04^{18}	9.17^{18}	8.90^{54}
Alignment Q	7.36^{18}	6.04^{18}	9.09^{18}	7.50^{54}
	7.93^{36}	7.54^{36}	9.13^{36}	$\mathbf{8.20^{108}}$

Both main effects in this interaction had significantly high variance ratios, suggesting that the varieties gave different yields (though a significant F does not mean that all *three* varieties differ from each other, hence the further t-tests we need, see Chapter 16), and that you get a higher leek yield by planting with the leaf fan parallel to rather than across the direction of the row. But we are warned against accepting such interpretations by the significant interaction! If you look within the black frame, you will see that, although variety C is the best yielder at both alignments, varieties A and B show the reverse superiority in yield in the two alignments. So which alignment the grower should pick is not clear-cut in the way the end totals would suggest; it depends on the variety chosen.

Before we leave this experiment and move on to Chapter 15, it is a long time since I've reminded you of what we are doing in using SqADS to calculate sums of squares. You may need to go back to page 95 to recall that SqADS is just a quick way on a calculator to sum all the squared differences between the data and the overall grand mean. So let's apply this thinking to the table of means for Variety × Alignment above.

In calculating the sum of squares for the figures in the body of the table, we assume that all 18 plots (3 replicates × 2 Fertilizers × 3 Trimmings) represented by each of the six means in the body of the table are identical. This gives us different 108 plot data from those actually obtained in the experiment (shown in the table at the start of this chapter). Our new table of 108 numbers assumes that all replicates per V × A mean have yielded the same, and that neither Fertilizers or Trimmings have had any effect whatsoever! So of our 108 numbers, 18 are all 8.50, another 18 are all 9.04, 18 are 9.17, and so on. Our sum of squares (remember page 15 that this is an abbreviation for *sum of squares of deviations from the mean*) of these 108 new figures is their squares added together of the difference between each figure and 8.20 (the overall mean). On a calculator (page 21) it is quicker to add the squared numbers (rather than their difference from 8.30) and subtract a correction factor based on the grand total (see page 22) – and SqADS of the 18-plot totals just gives us the same answer a lot more rapidly, by saving us repeatedly squaring the same numbers 18 times!

Addendum – additional working of sums of squares calculations

Variety $\times$ Alignment (V $\times$ A)

Interaction table for Variety $\times$ Alignment: 3 columns (Variety levels) $\times$ 2 rows (Alignment levels) = 6 supertotals

Divisor (number of plots per supertotal in body of table): 18

Number of treatment totals$_3$ per supertotal in body of table: $18/3 = 6$

Identification of first such supertotal: Treatments sharing A and P (dark shaded in table below). If we also shade (lighter) the totals contributing to the BQ and CP supertotals it is clear how the 36 treatment totals divide into the six supertotals of six figures we need.

A1PR	25.5_3	A2PR	27.0_3	B1PR	26.2_3	B2PR	27.6_3	C1PR	27.0_3	C2PR	28.8_3
A1PL	22.5_3	A2PL	23.4_3	B1PL	25.6_3	B2PL	28.5_3	C1PL	25.8_3	C2PL	27.0_3
A1PT	27.0_3	A2PT	27.6_3	B1PT	26.4_3	B2PT	28.4_3	C1PT	27.7_3	C2PT	28.7_3
A1Q	22.8_3	A2Q	22.5_3	B1QR	18.0_3	B2QR	18.6_3	C1QR	26.9_3	C2QR	28.2_3
A1QL	21.0_3	A2QL	21.6_3	B1QL	17.1_3	B2QL	17.9_3	C1QL	25.5_3	C2QL	27.0_3
A1QT	21.6_3	A2QT	22.9_3	B1QT	18.0_3	B2QT	19.2_3	C1QT	27.5_3	C2QT	28.6_3

Supertotals in interaction table:

	Variety A	Variety B	Variety C	
Alignment P	153.0_{18}	162.7_{18}	165.0_{18}	480.7_{54}
Alignment Q	132.4_{18}	108.8_{18}	163.7_{18}	404.9_{54}
	285.4_{36}	271.5_{36}	328.7_{36}	885.6_{108}

Table of sum of squares (SqADS):

Correction factor $= 7261.92$

Source of variation	Sum of squares		Source of variation	Sum of squares
Columns (Variety)	From earlier table	➡	Variety	49.44
Rows (Alignment)	From earlier table	➡	Alignment	53.20
Interaction	**by subtraction**		**Interaction**	**39.34**
Totals of table body	by SqADS	➡	Totals of table body	141.98

Variety × Trimming (V × T)

Interaction table for Variety × Trimming: 3 columns (Variety levels) × 3 rows (Trimming levels) = 9 supertotals

Divisor (number of plots per supertotal in body of table): 12

Number of treatment totals$_3$ per supertotal in body of table: 12/3 = 4

Identification of first such supertotal: Treatments sharing A and R (dark shaded in table below). If we also shade (lighter) the totals contributing to the AL and CT supertotals it is clear how the 36 treatment totals divide into the nine supertotals of four treatment totals we need (three coming from each third of the table):

A1PR	25.5_3	A2PR	27.0_3	B1PR	26.2_3	B2PR	27.6_3	C1PR	27.0_3	C2PR	28.8_3
A1PL	22.5^3	A2PL	23.4_3	B1PL	25.6_3	B2PL	28.5_3	C1PL	25.8_3	C2PL	27.0_3
A1PT	27.0_3	A2PT	27.6_3	B1PT	26.4_3	B2PT	28.4_3	C1PT	27.7_3	C2PT	28.7_3
A1QR	22.8_3	A2QR	22.5_3	B1QR	18.0_3	B2QR	18.6_3	C1QR	26.9_3	C2QR	28.2_3
A1QL	21.0_3	A2QL	21.6_3	B1QL	17.1_3	B2QL	17.9_3	C1QL	25.5_3	C2QL	27.0_3
A1QT	21.6_3	A2QT	22.9_3	B1QT	18.0_3	B2QT	19.2_3	C1QT	27.5_3	C2QT	28.6_3

Supertotals in interaction table:

	Variety A	Variety B	Variety C	
Trimming R	97.8_{12}	90.4_{12}	110.9_{12}	299.1_{36}
Trimming L	88.5_{12}	89.1_{12}	105.3_{12}	282.9_{36}
Trimming T	99.1_{12}	92.0_{12}	112.5_{12}	303.6_{36}
	285.4_{36}	271.5_{36}	328.7_{36}	885.6_{108}

Table of sum of squares (SqADS):

Correction factor = 7261.92

Source of variation	Sum of squares		Source of variation	Sum of squares
Columns (Variety)	From earlier table	➡	Variety	49.44
Rows (Trimming)	by SqADS		Trimming	6.59
Interaction	**by subtraction**		**Interaction**	**1.72**
Totals of table body	by SqADS	➡	Totals of table body	57.75

Fertilizer × Trimming (F × T)

Interaction table for Variety × Trimming: 2 columns (Fertilizer levels) × 3 rows (Trimming levels) = 6 supertotals

Divisor (number of plots per supertotal in body of table): 18

Number of treatment totals$_3$ per supertotal in body of table: 18/3 = 6

Identification of first such supertotal: Treatments sharing 1 and R (dark shaded in table below). If we also shade (lighter) the totals contributing to the 2L supertotals the pattern emerges of how the 36 treatment totals divide into the six supertotals of six treatment totals we need.

A1PR	25.5_3	A2PR	27.0_3	B1PR	26.2_3	B2PR	27.6_3	C1PR	27.0_3	C2PR	28.8_3
A1PL	22.5_3	A2PL	23.4_3	B1PL	25.6_3	B2PL	28.5_3	C1PL	25.8_3	C2PL	27.0_3
A1PT	27.0_3	A2PT	27.6_3	B1PT	26.4_3	B2PT	28.4_3	C1PT	27.7_3	C2PT	28.7_3
A1QR	22.8_3	A2QR	22.5_3	B1QR	18.0_3	B2QR	18.6_3	C1QR	26.9_3	C2QR	28.2_3
A1QL	21.0_3	A2QL	21.6_3	B1QL	17.1_3	B2QL	17.9_3	C1QL	25.5_3	C2QL	27.0_3
A1QT	21.6_3	A2QT	22.9_3	B1QT	18.0_3	B2QT	19.2_3	C1QT	27.5_3	C2QT	28.6_3

Supertotals in interaction table:

	Fertilizer 1	Fertilizer 2	
Trimming R	146.4_{18}	152.7_{18}	299.1_{36}
Trimming L	137.5_{18}	145.4_{18}	282.9_{36}
Trimming T	148.2_{18}	155.4_{18}	303.6_{36}
	432.1_{54}	453.5_{54}	$\mathbf{885.6_{108}}$

Table of sum of squares (SqADS):

Correction factor $= 7261.92$

Source of variation	Sum of squares		Source of variation	Sum of squares
Columns (Fertilizer)	From earlier table	➡	Fertilizer	4.24
Rows (Trimming)	From earlier table	➡	Trimming	6.59
Interaction	**by subtraction**		**Interaction**	**0.03**
Totals of table body	by SqADS	➡	Totals of table body	10.86

Alignment × *Trimming (A × T)*

Interaction table for Alignment × Trimming: 2 columns (Aligment levels) × 3 rows (Trimming levels) = 6 supertotals

Divisor (number of plots per supertotal in body of table): 18

Number of treatment totals$_3$ per supertotal in body of table: $18/3 = 6$

Identification of first such supertotal: Treatments sharing P and R (dark shaded in table below). It is clear that the six supertotals of six treatment totals we need are the sum of each of the six horizontal lines of the table:

A1PR	25.5_3	A2PR	27.0_3	B1PR	26.2_3	B2PR	27.6_3	C1PR	27.0_3	C2PR	28.8_3
A1PL	22.5_3	A2PL	23.4_3	B1PL	25.6_3	B2PL	28.5_3	C1PL	25.8_3	C2PL	27.0_3
A1PT	27.0_3	A2PT	27.6_3	B1PT	26.4_3	B2PT	28.4_3	C1PT	27.7_3	C2PT	28.7_3
A1QR	22.8_3	A2QR	22.5_3	B1QR	18.0_3	B2QR	18.6_3	C1QR	26.9_3	C2QR	28.2_3
A1QL	21.0_3	A2QL	21.6_3	B1QL	17.1_3	B2QL	17.9_3	C1QL	25.5_3	C2QL	27.0_3
A1QT	21.6_3	A2QT	22.9_3	B1QT	18.0_3	B2QT	19.2_3	C1QT	27.5_3	C2QT	28.6_3

Supertotals in interaction table:

	Alignment P	Alignment Q	
Trimming R	162.1_{18}	137.0_{18}	299.1_{36}
Trimming L	152.8_{18}	130.1_{18}	282.9_{36}
Trimming T	165.8_{18}	137.8_{18}	303.6_{36}
	480.7_{54}	404.9_{54}	$\mathbf{885.6_{108}}$

Table of sum of squares (SqADS):

Correction factor $= 7261.92$

Source of variation	Sum of squares		Source of variation	Sum of squares
Columns	From earlier table	⟹	Alignment	53.20
Rows (Trimming)	From earlier table	⟹	Trimming	6.59
Interaction	**by subtraction**		**Interaction**	**0.39**
Totals of table body	by SqADS	⟹	Totals of table body	60.18

Variety $\times$ *Fertilizer* $\times$ *Trimming (V $\times$ F $\times$ T)*

Interaction table for Variety $\times$ Fertilizer $\times$ Trimming: 3 columns (Variety levels) split into 2 (Fertilizer levels) $\times$ 3 rows (Trimming levels) $= 18$ supertotals

Divisor (number of plots per supertotal in body of table): $108/18 = 6$

Number of treatment totals$_3$ per supertotal in body of table: $6/3 = 2$

Identification of first such supertotal: Treatments sharing A, 1, and R (dark shaded in table below). By light shading A, 2, and L, and then B, 1, and T, we can see the pattern (three in each column) for locating the 18 supertotals of two treatment totals we need.

A1PR	**25.5$_3$**	A2PR	27.0$_3$	B1PR	26.2$_3$	B2PR	27.6$_3$	C1PR	27.0$_3$	C2PR	28.8$_3$
A1PL	22.5$_3$	*A2PL*	*23.4$_3$*	B1PL	25.6$_3$	B2PL	28.5$_3$	C1PL	25.8$_3$	C2PL	27.0$_3$
A1PT	27.0$_3$	A2PT	27.6$_3$	**B1PT**	**26.4$_3$**	B2PT	28.4$_3$	C1PT	27.7$_3$	C2PT	28.7$_3$
A1QR	**22.8$_3$**	A2QR	22.5$_3$	B1QR	18.0$_3$	B2QR	18.6$_3$	C1QR	26.9$_3$	C2QR	28.2$_3$
A1QL	21.0$_3$	*A2QL*	*21.6$_3$*	B1QL	17.1$_3$	B2QL	17.9$_3$	C1QL	25.5$_3$	C2QL	27.0$_3$
A1QT	21.6$_3$	A2QT	22.9$_3$	*B1QT*	*18.0$_3$*	B2QT	19.2$_3$	C1QT	27.5$_3$	C2QT	28.6$_3$

Supertotals in interaction table:

	Variety A		Variety B		Variety C		
	Fert. 1	Fert. 2	Fert. 1	Fert. 2	Fert. 1	Fert. 2	
Trimming R	48.3_6	49.5_6	44.2_6	46.2_6	53.9_6	57.0_6	299.1_{36}
Trimming L	43.5_9	45.0_9	42.7_6	46.4_6	51.3_6	54.0_6	282.9_{36}
Trimming T	48.6_9	50.5_9	44.4_6	47.6_6	55.2_6	57.3_6	303.6_{36}
	140.4_{18}	145.0_{18}	131.3_{18}	140.2_{18}	160.4_{18}	168.3_{18}	$\mathbf{885.6_{108}}$

Table of sum of squares (SqADS):

Correction factor $= 7261.92$

Source of variation	Sum of squares
Variety	49.44
Fertilizer	4.24
Trimming	6.59
V $\times$ F (remainder)	0.28
V $\times$ T (remainder)	1.72
F $\times$ T (remainder)	0.03
Interaction	**0.16**
Totals of table body	62.46

Variety $\times$ Alignment $\times$ Trimming (V $\times$ A $\times$ T)

Interaction table for Variety $\times$ Alignment $\times$ Trimming: 3 columns (Variety levels) split into 2 (Alignment levels) $\times$ 3 rows (Trimming levels) $= 18$ supertotals

Divisor (number of plots per supertotal in body of table): $108/18 = 6$

Number of treatment totals$_3$ per supertotal in body of table: $6/3 = 2$

Identification of first such supertotal: Treatments sharing A, P, and R (dark shaded in table below). By light shading A, P, and T, and then A, Q, and L, we can see the pattern (alternate rows across a pair of columns) for locating the 18 supertotals of two treatment totals we need.

A1PR	**25.5_3**	**A2PR**	**27.0_3**	B1PR	26.2_3	B2PR	27.6_3	C1PR	27.0_3	C2PR	28.8_3
A1PL	22.5_3	A2PL	23.4_3	B1PL	25.6_3	B2PL	28.5_3	C1PL	25.8_3	C2PL	27.0_3
A1PT	*27.0_3*	*A2PT*	*27.6_3*	B1PT	26.4_3	B2PT	28.4_3	C1PT	27.7_3	C2PT	28.7_3
A1QR	22.8_3	A2QR	22.5_3	B1QR	18.0_3	B2QR	18.6_3	C1QR	26.9_3	C2QR	28.2_3
A1QL	*21.0_3*	*A2QL*	*21.6_3*	B1QL	17.1_3	B2QL	17.9_3	C1QL	25.5_3	C2QL	27.0_3
A1QT	21.6_3	A2QT	22.9_3	B1QT	18.0_3	B2QT	19.2_3	C1QT	27.5_3	C2QT	28.6_3

Supertotals in interaction table:

	Variety A		Variety B		Variety C		
	Align. P	Align. Q	Align. P	Align. Q	Align. P	Align. Q	
Trimming R	52.5_6	45.3_6	53.8_6	36.6_6	55.8_6	55.1_6	299.1_{36}
Trimming L	45.9_9	42.6_9	54.1_6	35.0_6	52.8_6	52.5_6	282.9_{36}
Trimming T	54.6_9	44.5_9	54.8_6	37.2_6	56.4_6	56.1_6	303.6_{36}
	153.0_{18}	132.4_{18}	162.7_{18}	108.8_{18}	165.0_{18}	163.7_{18}	$\mathbf{885.6_{10}}$

8

Table of sum of squares (SqADS):

Correction factor $= 7261.92$

Source of variation	Sum of squares
Variety	49.44
Alignment	53.20
Trimming	6.59
F $\times$ A (remainder)	39.34
F $\times$ T (remainder)	1.72
A $\times$ T (remainder)	0.39
Interaction	**1.72**
Totals of table body	152.40

Fertilizer $\times$ Alignment $\times$ Trimming (F $\times$ A $\times$ T)

- **Interaction table for Fertilizer $\times$ Alignment $\times$ Trimming:** 2 columns (Fertilizer levels) split into 2 (Alignment levels) $\times$ 3 rows (Trimming levels) $=$ 12 supertotals
- **Divisor (number of plots per supertotal in body of table):** $108/12 = 9$
- **Number of treatment totals$_3$ per supertotal in body of table:** $6/3 = 3$
- **Identification of first such supertotal:** Treatments sharing 1, P, and R (dark shaded in table below). By light shading 2, P, and L, and then 1, P, and T, we can see the somewhat complicated pattern (alternate triplets along the rows) for locating the 12 supertotals of three treatment totals we need.

A1PR	**25.5$_3$**	A2PR	27.0$_3$	**B1PR**	**26.2$_3$**	B2PR	27.6$_3$	**C1PR**	**27.0$_3$**	C2PR	28.8$_3$
A1PL	22.5$_3$	**A2PL**	**23.4$_3$**	B1PL	25.6$_3$	**B2PL**	**28.5$_3$**	C1PL	25.8$_3$	**C2PL**	**27.0$_3$**
A1PT	**27.0$_3$**	A2PT	27.6$_3$	**B1PT**	**26.4$_3$**	B2PT	28.4$_3$	**C1PT**	**27.7$_3$**	C2PT	28.7$_3$
A1QR	22.8$_3$	A2QR	22.5$_3$	B1QR	18.0$_3$	B2QR	18.6$_3$	C1QR	26.9$_3$	C2QR	28.2$_3$
A1QL	21.0$_3$	A2QL	21.6$_3$	B1QL	17.1$_3$	B2QL	17.9$_3$	C1QL	25.5$_3$	C2QL	27.0$_3$
A1QT	21.6$_3$	A2QT	22.9$_3$	B1QT	18.0$_3$	B2QT	19.2$_3$	C1QT	27.5$_3$	C2QT	28.6$_3$

Supertotals in interaction table:

	Fertilizer 1		Fertilizer 2		
	Align. P	Align. Q	Align. P	Align. Q	
Trimming R	78.7$_9$	67.7$_9$	83.4$_9$	69.3$_9$	299.1$_{36}$
Trimming L	73.9$_9$	63.6$_9$	78.9$_9$	66.5$_9$	282.9$_{36}$
Trimming T	81.1$_9$	67.1$_9$	84.7$_9$	70.7$_9$	303.6$_{36}$
	233.7$_{27}$	198.4$_{27}$	247.0$_{27}$	206.5$_{27}$	**885.6**$_{108}$

Table of sum of squares (SqADS):

Correction factor = 7261.92

Source of variation	Sum of squares
Fertilizer	4.24
Alignment	53.20
Trimming	6.59
F × A (remainder)	0.25
F × T (remainder)	0.03
A × T (remainder)	0.39
Interaction	**0.14**
Totals of table body	64.84

Spare-time activity

The data relate to an experiment where three plant species were germinated with and without a fungicidal seed dressing in three soil types (there were three blocks). Per cent emergence was recorded.

The data would take up a lot of space and moreover I want to save you time on repetitive work. I've therefore done Phase 1 and 2 of the analysis for you, and this saves giving you the data for each block:

Source of variation	d.f.	sum of squares
Treatments	17	32,041.50
Blocks	2	356.77
Residual	34	3,199.40
TOTAL	**53**	**35,597.67**

Correction factor: = 264,180.17

Your assignment is to factorialize the 32,041.50 sum of squares and the 17 degrees of freedom for "Treatments." The table below is therefore a table of treatment totals, i.e. the sum of the three blocks for each treatment (hence the subscript 3) – who ever heard of 266% germination, anyway (!), for combining into supertotals.

	Silt loam	Sand	Clay
Petunia			
Untreated	266_3	286_3	66_3
Fungicide	276_3	271_3	215_3
Godetia			
Untreated	252_3	289_3	167_3
Fungicide	275_3	292_3	203_3
Clarkia			
Untreated	152_3	197_3	52_3
Fungicide	178_3	219_3	121_3

Then calculate mean squares and variance ratios. Just one of the 2-factor interactions is statistically significant. Which is it?

Keep your calculations. A further exercise in the interpretation of this experiment will be found in the "spare time activities" following Chapter 16, which explains how differences between means should be evaluated in a factorial analysis.

15

Factorial experiments with split plots

Chapter features

Introduction

The split plot design for factorials has special advantages for certain types of work. I will discuss later where these advantages stem from, but for the time being we can list the two most important as:

- When the application of one of the factors in the factorial would be impracticable on small plots – aerial spraying of herbicide would be a dramatic example!
- When one factor is only included because the interest of the experiment is in the interaction of that factor with a second factor. Here a good example is the use of insecticide to measure host plant resistance to a pest. We do not need to use the experiment to show that insecticides raise yields. A randomized block design would do this better; as will be made clear later, a split plot design involves a penalty in detecting the effect of one of the factors. In this experiment, the most resistant varieties will have few aphids, and so little or no protection with insecticide – therefore, the yields of sprayed and unsprayed plots will be similar. Therefore what is of interest in this experiment is not whether insecticides kill aphids, but how far sprayed and unsprayed plots differ in yield – in other words, the variety × insecticide interaction. That it would be hard to spray small plots without insecticide drifting where it is not wanted (see first advantage) makes a split plot design even more appropriate in this example!

Deriving the split plot design from the randomized block design

To derive the split plot design we'll begin with the kind of factorial experiment with which we are already familiar, i.e. laid out as randomized blocks. For our example, we'll use the experiment just mentioned – an aphid plant resistance trial of three Brussels sprout varieties (A, B, and C), sprayed (S) and unsprayed (U) with insecticide to detect the aphid-resistant variety, i.e. the one showing the smallest yield improvement when protected against aphid damage by spraying. Figure 15.1a shows how four replications of 3 varieties × 2 sprayings (=6 treatments per replicate) might be laid out as randomized blocks with 24 plots in total.

We can sacrifice detecting the obvious – that the insecticide will kill aphids on all varieties (as measured by yield) – if the benefit of doing this is that we get a more accurate discrimination of the interaction, i.e. how the effect of insecticide differs on the three varieties. Look at the left hand replicate. If the soil, or anything else affecting yield, varies within the replicate, such variation will probably interfere less with the contrast between adjacent plots (e.g. AU and AS) caused by the designed treatments than with the contrast between the more widely separated AU and CU. It usually works out in practice that (in the absence of any experimental treatments to make them different) adjacent plots are more similar than those further apart! Moreover, it would be hard to spray AS without insecticide drifting onto the adjacent unsprayed plots AU and BU.

To turn these 24 plots into a split plot design by easy stages, let's forget about spraying altogether and think how we would use the same plots for a straight comparison of the yield of three sprout varieties. Well, with 24 plots and three varieties, we could obviously step up the replication from 4 to 8, making it a much better variety experiment (Fig. 15.1b).

The allocation of degrees of freedom would be:

3 Varieties	2
8 Blocks	7
Residual	$2 \times 7 = 14$
Total (24 plots)	23

Now let's forget about varieties, and use the same area of land for just one variety, sprayed (S) and unsprayed (U) with insecticide, as four replicates of larger plots. We might well lay out the four replicates as shown in Fig. 15.1c.

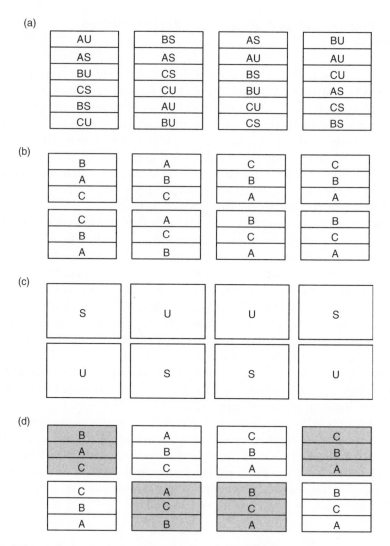

Fig. 15.1 Relationship between a randomized and a split plot design: (a) How the six factorial treatments of three Brussels sprout varieties (A, B, and C) either sprayed (S) or unsprayed (U) with insecticide might be allocated to four blocks. (b) How the same land area might be used for eight randomized blocks of just the three Brussels sprout varieties. (c) How the same land area might be used for four larger randomized blocks (arranged left to right across the diagram of the sprayed and unsprayed comparison. (d) The split plot design: each plot in design (c) – half a replicate of the sprayed/unsprayed comparison – is divided into three subplots as a full replicate of the variety comparison of design (b) by superimposing plans (b) and (c) on each other.

The allocation of degrees of freedom would be:

2 Sprayings	1
4 Blocks	3
Residual	$1 \times 3 = 3$
Total (8 plots)	7

Note that the insecticide experiment is less precise than the variety experiment for two reasons.

1 There are fewer replicates (blocks) contributing to fewer degrees of freedom as the divisor for residual variation.
2 The residual variation in the comparison of Sprayed and Unsprayed will anyway be larger as the distance on the ground within the replicates is greater than that within the replicates of the eight-replicate (block) variety layout.

We now have two experimental plans for the same site involving two experiments of different sizes – a four-replicate spraying experiment and an eight-replicate variety experiment.

The split plot experiment is nothing more complicated than doing both experiments on the same piece of land at the same time! In the plan (Fig. 15.1d), shaded plots are the ones that are sprayed (S).

You should perhaps now go back to the randomized block plan for this variety × insecticide experiment (Fig. 15.1a) and compare it with Fig. 15.1d to make sure you are clear how the split plot design differs. The basic concept is that the smaller plots of one factor (variety in this example) are "nested" within the larger plots of the other factor (insecticide). Indeed in the USA, split plot designs are usually known as "nested" designs – which I think is a better, more descriptive term than 'split plot."

Indeed, you can "nest" more than two factors. With three factors you would have not only "main plots" and "sub-plots," but also "sub-sub-plots"! This is akin to doing **three** experiments at the same time on the same plot of land. I think I'll leave it at that, though of course you could nest even more factors. Trying to sort out the analysis and degrees of freedom for a 4-factor split plot experiment is the sort of thing you can pass the time with if you're ever stuck at an airport for hours on end!

In the randomized block plan, the unit of area sprayed or unsprayed was the same size as the unit of area planted to a variety; therefore both treatments were replicated equally and the same residual sum of squares is appropriate for obtaining the variance ratio for both comparisons. The whole purpose of randomization is that positional variation between the plots in any one replicate should equally affect both spraying and variety comparisons.

In the split plot design, by contrast, we actually want a different residual mean square for each factorialized treatment. The sprayed/unsprayed comparison will probably have a larger residual sum of squares since the unit of area treated is larger and thus encompasses more positional variability. Furthermore, the fewer replicates for spraying than for varieties means that the residual sum of squares will then have a smaller divisor (the fewer degrees of freedom). Thus a split plot analysis of variance involves calculating **two** residual sums of squares, one for the four-replicate spraying experiment and a different one for the eight-replicate variety experiment.

What about the spraying × variety interaction, the real purpose of the experiment? Well, the beauty of the split plot design is that the interaction involves the eight-replicate experiment and will therefore be tested against the potentially smaller residual mean square, thus raising the chances of detecting the interaction with a higher variance ratio (F value).

Degrees of freedom in a split plot analysis

Main plots

Potential confusion here! The term "main plots" refers to the larger plots (for the sprayed/unsprayed comparison in our example) which are split into "sub-plots" for the second factor (varieties). Yet in terms of what we hope for from the experiment, it is the sub-plot comparisons which are of "main" importance, particularly the spraying × variety interaction which the sub-plot part of the analysis includes. So the "main" plots paradoxically carry the **minor** interest!

We have already (page 197) set out the degrees of freedom for the main plot part of the experiment as:

2 Spraying treatments	1 d.f.
4 Blocks (replicates)	3 d.f.
Residual	$1 \times 3 = 3$ d.f.
Total (8 main plots)	7 d.f.

Sub-plots

These are the smaller plots in the part of experiment with greater replication, and this part of the analysis is where the greatest interest lies.

Nothing can alter the fact that there are 24 plot in the whole split plot experiment. This therefore gives 23 d.f. Of these 23 d.f., 7 have just been allocated to the main plots, which must leave 16 for the sub-plots. But when we try and allocate the 16 degrees of freedom, it just doesn't add up! With 2 d.f. for three varieties and 7 for eight replicates, all our previous experience tells us the residual must be $2 \times 7 = 14$ d.f. But then this would add up to more than $16 - 23$ in fact!

3 Varieties	2 d.f.
8 Blocks (replicates)	7 d.f.
Residual	$2 \times 7 = 14$ d.f.
Total for sub plots	16 d.f.

Don't read on just yet, but first look back to the experimental plan (Fig. 15.1d) and try and work out for yourself how the discrepancy has arisen. I'll be really proud of you if you can!

OK – did you make it? The table above would add up to 16 d.f. if we deleted the 7 d.f. for the eight blocks, wouldn't it? And that is exactly what we should do. The 7 degrees of freedom are from the eight replicates, which themselves are the eight plots of the main plot experiment. So the 7 d.f. for the total variation of the main plot part of the analysis is the same d.f. and variation as associated with the eight blocks of the sub-plot part of the analysis. So these 7 d.f. in the sub-plot allocation have already been allocated! – in the main plot allocation of d.f.

So now the sub-plot d.f. add up to the required 16, but I fear the allocation of d.f. in that part of the experiment is still not right. Remember (page 198) that the spraying × variety interaction is part of the sub-plot analysis, so we have still to fit in the 2 d.f. (from 1 d.f. for spraying × 2 d.f. for varieties). If we do this, the residual now drops to 12:

3 Varieties	2 d.f.
2 spraying × 3 Variety interaction	$1 \times 2 = 2$ d.f.
Residual	$(2 \times 7) - 2 = 12$ d.f.
Total	16 d.f.

But why is this right? It's because the residual of an equivalent randomized block experiment of 2 sprayings × 3 varieties = 6 treatments with 4 replicates (i.e. still 24 plots with 23 total d.f.) would be $5 \times 3 = 15$ d.f. In the split plot analysis 3 of these 15 d.f. are allocated to the residual of the main plot part of the experiment, correctly leaving $15 - 3 = 12$ for the sub-plot residual.

It may help if I show how the degrees of freedom for a randomized block factorial jump around to give the allocation of d.f. for a split plot analysis. Both experiments have 24 plots, and involve the same factorialized treatments, 2 sprayings and 3 varieties:

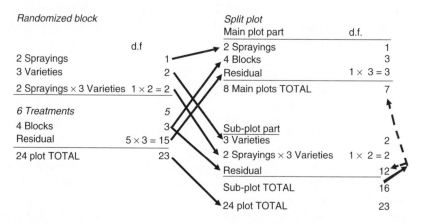

The allocation of degrees of freedom and calculation of sums of squares for a split plot analysis can be confusing compared with a randomized block factorial design as, since there is no single undivided sum of squares for "Treatments" there cannot be the familiar Phase 2, nor therefore can there be a Phase 3 splitting up the "treatment" variation into its components. Analysing a split plot experiment therefore has to start from scratch.

The secret is to do the calculations in the right order. The following table therefore repeats the allocation of degrees of freedom for our 2-factor split plot, but has added to it the most helpful order of the calculations, which will apply to any 2-factor split plot experiment:

Main plot part	d.f.	Order
2 Sprayings	1	3
4 Blocks	3	4
Residual	$1 \times 3 = 3$	5
8 Main plots TOTAL		2
Sub-plot part		
3 Varieties	2	7
2 Sprayings $\times$ 3 Varieties	$1 \times 2 = 2$	8
Residual	12	9
Sub-plot TOTAL	16	6
24 plot TOTAL	23	1

B	25.5		A	21.1		C	28.6		C	29.3
A	24.9		B	17.9		B	19.5		B	29.2
C	25.8		C	28.9		A	23.2		A	29.8

C	26.1		A	27.6		B	29.7		B	21.3
B	18.0		C	29.4		C	30.3		C	31.0
A	21.7		B	29.3		A	29.5		A	25.8

Fig. 15.2 Yield of sprouts (kg/plot) from the experimental design shown in Fig. 15.1d.

Numerical example of a split plot experiment and its analysis

We might as well use the yield data (Fig. 15.2) from the experiment on spraying and Brussels sprout varieties which has formed the basis of this chapter so far. As before the shaded areas have been sprayed and A, B, and C refers to the three sprout varieties. Final yield of sprouts was measured in kg/plot.

As in previous analyses, we first have to unscramble the randomization to gain easy access to the totals we will need for **SqADS**. Note that this table of data would look just the same had the layout been randomized blocks; we cannot tell the design from the table. As before, we will use subscripts to track how many plots contribute to a total:

Table A

Replicate	SA	SB	SC	UA	UB	UC	Row totals
1	24.9	25.5	25.8	21.7	18.0	26.1	142.0_6
2	27.6	29.3	29.4	21.1	17.9	28.9	154.2_6
3	29.5	29.7	30.3	23.2	19.5	28.6	160.8_6
4	29.8	29.2	29.3	25.8	21.3	31.0	166.4_6
Column totals	111.8_4	113.7_4	114.8_4	91.8_4	76.7_4	114.6_4	$\mathbf{623.4_{24}}$

This table is the data for the 24 sub-plots. We now need a second table for the eight main plots, which are the four replicates of areas sprayed or unsprayed. Each line of data on either side of the double vertical line is the three variety sub-plots in one sprayed (S) or unsprayed (U) main plot. So to get the eight totals for the main plots, we add the three numbers in each replicate on either side of the double vertical line (e.g. $24.9 + 25.5 + 25.8 = 76.2_3$) to get:

Table B

Replicate	Sprayed	Unsprayed	Row totals
1	76.2_3	65.8_3	142.0_6
2	86.3_3	67.9_3	154.2_6
3	89.5_3	71.3_3	160.8_6
4	88.3_3	78.1_3	166.4_6
Column totals	340.3_{12}	283.1_{12}	$\mathbf{623.4_{24}}$

Calculating the sums of squares

OK, so we can start "number crunching," and we begin with calculating the correction factor for all the sums of squares as the (Grand Total)$^2/24 = 623.4^2/24 = 16{,}192.82$. This correction factor applies regardless of whether a sum of squares is in the main or sub-plot part of the analysis.

In calculating sums of squares we will follow the order suggested in the earlier table of the allocated degrees of freedom. To save repetition of the detail of the calculations, the relevant numbered paragraph will be asterisked if such details are given in Box 15.1 (page 203). The number of totals to square and add × the divisor will always be the number of plots $= 24$ in this experiment.

But before we start calculating sums of squares, it is worth drafting an interaction table for sprayings × varieties, using the format we've used previously (page 159):

Table C

Varieties

Sprayings		A	B	C	
	S	111.8_4	113.7_4	114.8_4	340.3_{12}
	U	91.8_4	76.7_4	114.6_4	283.1_{12}
		203.6_8	190.4_8	229.4_8	GT $= \mathbf{623.4_{24}}$

So off we go, calculating the sums of squares in the order suggested. It will be a good exercise for you, as you reach each step in the sequence, to identify the numbers you will need for that step. They can be found in

BOX 15.1

1 The total sum of squares involves adding the squares of all 24 data in Table A and subtracting the correction factor ($623.4^2/24$), *viz.*:

$$24.9^2 + 27.6^2 + 29.5^2 + 29.8^2 + 25.5^2 + \cdots + 31.0^2 - 16{,}192.82$$
$$= 391.00$$

2 The number at the start of the relevant line in the analysis table clues us in that there are eight main plot totals for SqADS, with the divisor of 3 since $24/8 = 3$. Each main plot is one replicate of three varieties, so with six plots per replicate there are two main plot totals per replicate. There they are, in the body of Table B:

$$\frac{76.2^2 + 86.3^2 + 89.5^2 + 88.3^2 + 65.8^2 + \cdots + 78.1^2}{3}$$
$$- 16{,}192.82 = 202.85$$

3 There are supertotals of 12 plots each for the two levels of Sprayings, S and U.
They are the two row totals in Table C:

$$\frac{340.3^2 + 283.1^2}{12} - 16{,}192.82 = 136.32$$

4 Four replicate totals (row totals in Table A) must have a divisor of 6:

$$\frac{142.0^2 + 154.2^2 + 160.8 + 166.4^2}{6} - 16{,}192.82 = 202.85$$

5 Three Variety supertotals (column totals in Table C) are each totals of eight plots (8 is therefore the divisor for SqADS):

$$\frac{203.6^2 + 190.4^2 + 229.4^2}{8} - 16{,}192.82 = 98.37$$

The interaction. 2 Sprayings $\times$ 3 Varieties gives six totals (each of $24/6 = 4$ plots) – these are six totals with subscript 4 inside the black frame of Table C. But remember that interactions are always found as remainders – in this case once the sums of squares already calculated for Sprayings (*136.32*) and Varieties (*98.37*) have been deducted!

$$\frac{111.8^2 + 113.7^2 + 114.8^2 + 91.8^2 + 76.7^2 + 114.6^2}{4}$$
$$- 16{,}192.82 - 136.32 - 98.37 = 84.81$$

either Table A, B, or C – and there's always Box 15.1 to help if you're in trouble.

1 *Total sum of squares.* We square and add all 24 plot data in Table A and subtract the correction factor to get 391.00.

2 *Main plot sum of squares.* $\mathbf{SqA}^{\text{8 Main plot totals}}\mathbf{D}^{\text{by 3}}\mathbf{S}^{\text{Correction Factor}}$. Eight totals, each totals of three plots? These are the ones in the body of Table B. The sum of squares is 202.05. The next three steps involve the same table and partition this total main plot sum of squares into its components.

3 *Sprayings sum of squares.* $\mathbf{SqA}^{\text{The 2 S and U totals}}\mathbf{D}^{\text{by 12}}\mathbf{S}^{\text{Correction Factor}}$. Two totals, each totals of 12 plots? These have to be the row supertotals of Table C, 340.3 and 283.1. SqADS gives us 136.32.

4 *Replicates sum of squares.* $\mathbf{SqA}^{\text{4 Replicate totals}}\mathbf{D}^{\text{by 6}}\mathbf{S}^{\text{Correction Factor}}$. The four totals we need are the four row totals in Tables A or B. SqADS gives us 55.05.

5 *Main plot residual sum of squares.* This remainder from the main plot sum of squares is found by subtraction, and similarly . . .

6 Subtraction of the main plot sum of squares from the total sum of squares gives the sub-plot sum of squares. We can now begin to fill in the analysis of variance table:

Lead line: 2 Sprayings × 3 Varieties (= 6 treatments) ×
4 Replicates = 24 plots
Correction factor: 16,192.82

Source of variation	Degrees of freedom	Sum of squares	*Order of calculation*	Mean square	Variance ratio	*P*
Main plots						
2 Sprayings	1	136.32	*3*			
4 Replicates	3	55.05	*4*			
Main plot residual	3	10.68	*5*			
8 Main plots total	**7**	**202.05**	*2*			
Sub-plots						
3 Varieties	2		*7*			
2 Sprayings × 3 Varieties	2		*8*			
Sub-plot residual	12		*9*			
Sub-plot total	**16**	**188.95**	*6*			
TOTAL (24 plots)	23	391.00	*1*			

7 *Varieties sums of squares.* $\mathbf{SqA}$ ^3 Variety totals^ $\mathbf{D}$^by 8^ $\mathbf{S}$^Correction Factor^. The three variety supertotals (each of eight plots) are easy to identify as the column totals of Table C – 203.6, 190.4, and 229.4. SqADS gives us 98.37.

8 *Spraying × Varieties interaction sum of squares.* $\mathbf{SqA}$^2×3=6 Spraying×Varieties totals^ $\mathbf{D}$^by 4^ $\mathbf{S}$^Correction Factor^. The six interaction totals are the "treatment" or column totals of Table A, also the six totals in the body of Table C. SqADS gives us 319.50. Can that be right? It's larger than the entire sub-plot total of 188.95. If this puzzles you, then I fear you have forgotten something very important (page 152).

The sum of squares for an interaction can never be calculated directly, it is always a remainder – what is left over when the sums of squares of contributing sources of variation (in this case Sprayings and Varieties have been deducted). **Interactions are always remainders.** Surely that rings a bell? So we have to subtract the 136.32 for Sprayings and the 98.37 for Varieties to leave 84.81 as the remaining sum of squares attributable to the interaction.

9 The last figure, the *sub-plot residual sum of squares*, is found by subtraction.

End Phase

We can now complete the END PHASE by calculating mean squares and variance ratios. Note that we have two residual sum of squares (3.55 and 0.48), and each is used for the variance ratios in its part of the analysis.

Lead line: 2 Sprayings × 3 Varieties (= 6 treatments) × 4 Blocks (replicates) = 24 plots

Correction factor: 16,192.82

Source of variation	Degrees of freedom	Sum of squares	Mean square	Variance ratio	P
Main plots					
2 Sprayings	1	136.32 ──────→	136.32 ─┐ ┌→	38.29	<0.05
4 Replicates	3	55.05 ──────→	16.35 ↑ ├→	5.15	ns
Main plot residual	3	10.68 ──────→	3.55 ↑		
Main plot total	7	202.05			
Sub-plots					
3 Varieties	2	98.37 ──────→	49.19 ─┐→	102.47	<0.001
2 Sprayings × 3 Varieties	2	84.81 ──────→	42.41 ↑ ├→	88.34	<0.001
Sub-plot residual.	12	5.77 ──────→	0.48 ↑		
Sub-plot total	16	188.95			
TOTAL (24 plots)	23	391.00			

The Sprayings × Varieties interaction is highly significant, indicating that the sprout varieties differ in how far insecticide is necessary to protect the yield from aphid damage. This difference between the varieties is the main phenomenon being tested in this experiment. Even without doing the additional tests on differences between means described in the next chapter, the totals in the body of Table C above show that Variety C is resistant to aphids compared with Varieties A and B since it shows so little yield difference when protected from aphids by spraying.

Comparison of split plot and randomized block experiment

It would be nice to be able to compare the above analysis with that for the randomized block design shown at the start of this chapter (page 196), since both designs have 24 plots in a matrix of four columns of six plots. But to mean anything, we would have to do the two different experiments on the same piece of land at the same time – impossible, unless we do it virtually! That's the best we can do, and how we play with the data we have from the split plot to get a best guess of the results we would have got from the randomized block design is explained in Box 15.2. You can skip this and move straight to the unscrambled table of data below, but Box 15.2 is perhaps a useful demonstration of how we can use the principle of the analysis of variance for another purpose. The "virtual" results are as follows:

Replicate	SA	SB	SC	UA	UB	UC	Row totals
1	24.9	27.2	26.1	20.1	16.3	27.4	142.0_6
2	26.7	26.5	28.9	23.7	20.1	28.3	154.2_6
3	27.9	28.6	30.2	23.3	20.5	30.3	160.8_6
4	30.1	31.2	30.0	23.8	19.8	30.5	165.4_6
Column totals	109.6_4	113.5_4	115.2_4	90.9_4	76.7_4	116.5_4	$\mathbf{622.4_{24}}$

Compare this with the analogous table for the split plot experiment (page 201). It is not possible to tell from the two tables that they reflect different experimental designs. How a data table is analyzed is determined by the plan of the experiment, not from the table of results!

I trust it's not necessary to give the detail of analyzing the above table as a randomized block experiment, but it might be a good spare-time activity

BOX 15.2

Calculations from the split plot data

We begin with the six "treatment" means of the four replicates: SA = 28.0, SB = 28.4, SC = 28.7, UA = 23.0, UB = 19.2, UC = 28.7. If our "treatments" were the only variation in the data, then all plots per "treatment" would show the same value. But positional effects (both within and between replicates) leads to other variation, measured by the difference between the "treatment" mean and the observed datum for the plot. This is shown on the plan below by the treatment mean underlined, then the plot datum, and finally the difference in bold. As before A, B, and C are the three varieties, and the sprayed plots are shaded:

B 28.4, 25.5, **–2.9**	A 23.0, 21.1, **–1.9**	C 28.7, 28.6, **–0.1**	C 28.7, 29.3, **0.6**
A 28.0, 24.9, **–3.1**	B 19.2, 17.9, **–1.3**	B 19.2,19.5, **0.3**	B 28.4, 29.2, **0.8**
C 28.7, 25.8, **–2.9**	C 28.7, 28.9, **0.2**	A 23.0, 23.2, **0.2**	A 28.0, 29.8, **1.8**

C 28.7, 26.1, **–2.6**	A 28.0, 27.6, **–0.4**	B 28.4, 29.7, **1.3**	B 19.2, 21.3, **2.1**
B 19.2, 18.0, **–1.2**	C 28.7, 29.4, **0.7**	C 28.7, 30.3, **1.6**	C 28.7, 31.0, **2.3**
A 23.0, 21.7, **–1.3**	B 28.4, 29.3, **0.9**	A 28.0, 29.5, **1.5**	A 23.0, 25.8, **2.8**

If we now remove the treatment means and plot data from the above layout, there is left (in bold typeface) how yield would change if no treatments had been applied and we were just measuring uniformity of yield across the site. A gradient of increased yield from left to right is clear. Here are these figures in bold on their own:

–2.9	–1.9	–0.1	0.6
–3.1	–1.3	0.3	0.8
–2.9	0.2	0.2	1.8
–2.6	–0.4	1.3	2.1
–1.2	0.7	1.6	2.3
–1.3	0.9	1.5	2.8

(Continued)

BOX 15.2 Continued

Converting to randomized blocks

A <u>23.0</u>, **-2.9**, *20.1*	B <u>28.4</u>, **-1.9**, *26.5*	A <u>28.0</u>, **-0.1**, *27.9*	B <u>19.2</u>, **0.6**, *19.8*
A <u>28.0</u>, **-3.1**, *24.9*	A <u>28.0</u>, **-1.3**, *26.7*	A <u>23.0</u>, **0.3**, *23.3*	A <u>23.0</u>, **0.8**, *23.8*
B <u>19.2</u>, **-2.9**, *16.3*	C <u>28.7</u>, **0.2**, *28.9*	B <u>28.4</u>, **0.2**, *28.6*	C <u>28.7</u>, **1.8**, *30.5*
C <u>28.7</u>, **-2.6**, *26.1*	C <u>28.7</u>, **-0.4**, *28.3*	B <u>19.2</u>, **1.3**, *20.5*	A <u>28.0</u>, **2.1**, *30.1*
B <u>28.4</u>, **-1.2**, *27.2*	A <u>23.0</u>, **0.7**, *23.7*	C <u>28.7</u>, **1.6**, *30.3*	C <u>28.7</u>, **2.3**, *30.0*
C <u>28.7</u>, **-1.3**, *27.4*	B <u>19.2</u>, **0.9**, *20.1*	C <u>28.7</u>, **1.5**, *30.2*	B <u>28.4</u>, **2.8**, *31.2*

Now we re-allocate the experimental treatments to the plots according to the plan for randomized blocks (Fig. 15.1a). Each plot now has (from left to right) underlined the treatment mean taken from the split plot data, e.g. mean for variety A when sprayed = 28, the positional variation in bold (from the table above) and, finally in italics the sum of the two, which becomes the hypothetical new datum for that plot. Again, the sprayed plots are shaded: The yields in italics are our estimate of what the results of the randomized block experiment would have been.

---

to do the calculations and check that you reach the same analysis table as this:

Lead line: 2 Sprayings × 3 Varieties (= 6 treatments) × 4 Blocks (Replicates) = 24 plots

Correction factor: 16,140.91

Source of variation	Degrees of freedom	Sum of squares	Mean square	Variance ratio	P
2 Sprayings	1	122.41	122.41	133.44	<0.001
3 Varieties	2	116.74	58.37	63.63	<0.001
2 Sprayings × 3 Varieties	2	90.80	45.40	49.49	<0.001
6 Treatments	*5*	*329.95*	*65.99*	*71.94*	*<0.001*
4 Replicates	3	51.67	17.22	18.78	<0.001
Residual	15	13.76	0.92		
TOTAL (18 plots)	23	395.38			

The sum of squares for residual variation at 13.76 is not that different from 16.45, the sum of the two residuals in the split plot table. In the randomized block analysis, this gives a single "averaged" mean square of 0.92 for calculating all the variance ratios. In the split plot analysis, the larger main plots show more positional variation between them than the small adjacent sub-plots. Indeed the residual sum of squares is split very unequally, with two-thirds coming from main plots. As the main plot residual has fewer degrees of freedom than the sub-plots, the residual mean square is four times larger (3.55) than the "averaged" 0.92 above, while that for the sub-plots is about halved (0.48). Hence, comparing a split plot design with a randomized block, the chances of detecting statistically significant effects of the main plot factor(s) decrease, while they increase for the sub-plot factors. Remember that this experiment aimed to measure plant susceptibility to aphids by the yield increase of sprayed over unsprayed plants; i.e. the interaction between the factors and part of the sub-plot part of the experiment.

Don't pay too much attention to the significance levels in the analysis tables. In this particular experiment, the randomized block has succeeded perfectly well in associating a high level of significance ($P < 0.001$) with the sub-plot components. The relative efficiency of the two parts of the split plot design compared with the randomized block is clearly seen if we compare the variance ratios (F values):

Source in the split plot design	F (split plot)	F (randomized block)
Main plot		
Sprayings	38.29 ⟶	133.44
Replicates	5.15 ⟶	18.78
Sub-plot		
Varieties	102.47 ⟵	63.33
Spraying × varieties interaction	88.34 ⟵	49.49

Look at the arrows. The shift in F values gives the split plot design an advantage for the sub-plot at the expense of the main plot factors. We now discuss why this "trade-off" between the two parts of the split-plot design might be beneficial.

Uses of split plot designs

Some uses were already alluded to at the start of this chapter, but it makes sense to discuss the uses more fully now that we've been through the features of the analysis.

1 The design can improve the detection of significant differences where some parts of an experiment (particularly interactions) are of particular

interest, and the overall affect of one of the factors is already known or of limited interest. Many people set all experiments out as randomized blocks, as if there were no other designs! If an experiment is factorial in nature, it will really be quite often that the aim is more specific than appropriate for randomized blocks – e.g. is it necessary to prove yet again that plants grow better with nitrogen fertilizer? Nitrogen fertilizer has probably only been included in the experiment because of interest in how it interacts with another factor (see first example below). Here are just three examples of experiments suitable for a split-plot design. I'm sure you'll pick up the general idea:

Does the yield enhancement obtained by removing the senescing lower leaves of tomato depend on the amount of N fertilizer applied? (N levels would be the main plot treatment; retaining or removing the senescing leaves would be the sub-plot treatment.)

Are some formulations of a particular pesticide more rain-fast than others? (Levels of simulated rainfall would be the main plot factor – that rainfall washes pesticide off plants can be assumed; different formulations would be the sub-plot factor.)

Is there an optimum timing for applying fruit thinner to apples, and is it the same for all varieties. Although we might be interested in overall yield differences between varieties, a split plot design would give us the best chance of detecting the optimum timing for the fruit thinner and the consistency between the varieties in relation to this timing. We might therefore be wise to make varieties the main plot factor and the fruit thinner timings the levels of the sub-plot factor.

2 Practical considerations may attract us to a split plot design. Land prepa-ration and many agronomic practices such as mechanical weeding may be hard to implement on a small-sized plot. A silly example was given near the start of the chapter – imagine spraying a 1 m square plot from the air without contaminating adjacent plots!

3 Here's a cunning use of a split plot design – how to get something for nothing! As soon as an experiment has a number of randomized blocks, we can assign whole blocks to different levels of another factor as "main plots" and so get "free" information about the effect of this factor. We may have no interest in the interaction between the two factors, the use of the main plot factor may just be a pilot study to see if it is worth experimenting in that direction.

Suppose a grower is doing a lettuce variety trial in a corner of a large glasshouse to find the most suitable variety for a winter crop. The grower also knows there are some soil-borne pathogens in his glasshouse soil, and wonders whether this has reached the stage where it would benefit him to splash out on soil fumigation for the whole glasshouse. Doing the lettuce variety trial gives him the chance to test soil fumigation out on a small scale as "main plots" superimposed on the lettuce variety trial. If there really is a beneficial effect on yield, he should be able to pick this up.

4 Sometimes it is obvious that the averaging of residual variation for the mean square would be illogical in relation to all the components of a factorial experiment. This usually arises where a factor cannot be randomized at the same level as the others. We may have too few plants to have different plants for measuring disease incidence on young, medium and old leaves, and have to use strata within single plants as one of our factors. In experiments with lighting treatments for plants, the inability to provide a different and randomized lighting treatment for every plant may force us to bulk plants given other treatments into main plots for a limited number of specific lighting conditions, even though such lighting conditions might be the factor of primary interest in the experiments.

Or we may be measuring growth effects, and need to take measurements in time. If we cannot make lots of multiple sowings at different times, we wouldn't be able to randomize plant age, and have to sample from plants growing from a single sowing. As a "better than nothing" solution, such data (with their very different residual variability) can be treated as a sort of split plot, with leaf strata or time as main plots. However, a more advanced statistical text should be consulted for the more complex analysis and significance tests involved.

Spare-time activity

The data are from an experiment comparing the productivity of Maris Bard potatoes grown from small (S = 40–60 g) or large (L = 90–110 g) tubers. There were 16 replicate rows with the halves of each row allocated at random to the two tuber sizes (the sub-plots). The tubers in pairs of adjacent rows were planted to be just covered with soil and then either immediately had the soil between the rows drawn up to make a ridge 9 cm high over the tubers (R) or not ridged until the plants were 10–15 cm high (F) (the main plots).

At harvest, for each sub-plot, assessments were of the number of tubers and their total fresh weight, and the fresh weight per tuber. Only the data for number of tubers are presented here:

Replicates	R		F	
	S	L	S	L
1	64	61	36	59
2	51	59	42	58
3	44	58	37	51
4	59	69	45	64
5	56	78	33	56
6	50	61	42	50
7	59	84	57	70
8	51	67	43	64

Complete a split-plot analysis of variance.

Is the interaction between planting technique and tuber size statistically significant?

Keep your calculations and return to the interpretation of the experiment as an exercise following Chapter 16.

16

The *t*-test in the analysis of variance

Chapter features

Introduction

This is the chapter which has been mentioned throughout the previous chapters on the analysis of variance as providing guidance to a fuller interpretation of the results than can be derived from the variance ratio (F) alone. At the end of an analysis of variance of plot yields of six varieties of a crop we may have been able to show that the variance ratio (F value) for varieties is significantly large. What does this mean? On page 107 I have already emphasized that it does no more than indicate that at least one of the varieties has a different yield from one of the others, from some of the others or from all the others. Even with a very high F significant at $P < 0.001$, it could be there is just one lousy or superb variety among the six, with the other five having essentially similar yields.

If there are only two varieties in the experiment, then a significantly high F is enough for us to accept that their yields are statistically different. However, when we have **more than two** means to compare, we – perhaps paradoxically – have to return to a test which can test differences **only** between **two** means, namely the *t*-test.

This test is a long way back, and you may wish either to turn back to Chapter 8 to refresh your memory or manage with the recap below.

Brief recap of relevant earlier sections of this book

Pages 58–69. With the t-test, we can evaluate the statistical significance of a difference between two means using the yardstick of the *standard error of differences between two means* (s.e.d.m.). Does this ring a bell?

$$\text{The } t\text{-test takes the form: is } \frac{\text{mean } x \text{ and mean } y \text{ difference}}{\text{s.e.d.m.}}$$

$$> \text{tabulated } t \text{ (usually for } P = 0.05)$$

If this is familiar, how about the way we calculate the s.e.d.m. (it's on page 51)? Do you remember that we got there by beginning with the variance (s^2) of a set of numbers? We had a mnemonic (page 52) for s.e.d.m. – "**S**quare root of the **E**stimated variance, after **D**oubling and then **M**ean-ing." Thus, in notation, the progression from variance to s.e.d.m. can be represented as:

$$s^2 \rightarrow 2s^2 \rightarrow 2s^2/n \rightarrow \sqrt{(2s^2/n)}$$
$$\text{E} \quad \text{D} \quad \text{M} \quad \quad \text{then } \text{S}$$

This sequence represents the steps: variance of individual numbers → variance of differences → variance of differences between two means of n numbers → standard error of differences between two means (s.e.d.m.) of n numbers.

Note: Where the two means are based on different numbers of replicates (see page 68 and page 81), we simply modify the calculation of the s.e.d.m. by using the algebraic identity that $\sqrt{2s^2/n}$ can be re-written as $\sqrt{s^2(1/n + 1/n)}$ or $s\sqrt{1/n + 1/n}$, giving us the opportunity to insert two different values of n. Simple!

Pages 107–108. How does $\sqrt{2s^2/n}$ relate to the analysis of variance (see page 107)? Well the "n" is the number of observations involved in the means being compared – that's fairly straightforward. What, however, is the variance (s^2)? It is not the total variance of the experiment, as that includes not only the random background variation (the s^2 we are looking for) but also variability that creates the difference between the means we want to test, i.e. the systematic variation caused by the experimental treatments and replicates. We have to subtract this systematic variation from the total variation to get at the s^2 appropriate for evaluating differences between the means. The s^2 we want is the measure of the random background variation and can be identified as *the residual variation* since, after all, this is what is left of the total sum of squares as a *remainder* after the sums of squares for treatments and replicates have been subtracted (page 89, if you need a reminder).

Well, residual variance is sums of squares divided by degrees of freedom; therefore it is what we have been calling the *residual mean square* in our analysis of variance tables. So that's the s^2 for our $\sqrt{2s^2/n}$ (or s.e.d.m.) calculation, making the s.e.d.m. $= \sqrt{(2 \times residual\ mean\ square/n)}$.

We could then simply do our *t*-tests between pairs of means from the experiment by asking:

$$\text{is } \frac{\text{mean A and mean B difference}}{\text{s.e.d.m.}} > \text{tabulated } t$$

(for $P = 0.05$ and the residual d.f.)?

Least significant difference test

A simple *t*-test is tedious with several means to test, and on page 108 I already introduced the concept of the *least significant difference (LSD)*, where we re-arrange the expression above to work out the minimum difference between means that would exceed the tabulated *t*. All differences larger than this would be "significant" at $P < 0.05$ by the *t*-test. We can re-arrange the *t*-test formula to get the LSD by solving for that mean A – mean B value which equals $t \times$ the s.e.d.m., i.e.

$$\text{LSD} = t_{(P=0.05,\ \text{residual d.f.}) \times \text{s.e.d.m.}}$$

or

$$\text{LSD} = t_{(P=0.05,\ \text{residual d.f.})} \times \sqrt{2 \times \text{residual mean square}/n}.$$

Pages 109–110. Here I pointed out that the *t*-test or LSD test in the above form fails to address the problem of comparisons between several means, such as say the yields of seven varieties of lettuce. There is no order in which we can expect these means to be arranged; each is totally independent of the others. Look again at Fig. 10.2 and Box 10.8 – seven means involve 21 tests if we compare each against all the others. On page 109 I illustrated this danger with just the first seven means of three eggs drawn from the single population of eggs featured way back on page 26. Yes, the largest and smallest of these seven means differed by more than the LSD (at $P = 0.05$, the 1 in 20 chance) even though we know that the conclusion that the two means were drawn from different populations of eggs is false.

Sometimes, in relation to means from an experiment, people ask "If I pretend I had done the experiment with only the two treatments with which I want to compare the means, I could then just use my LSD, couldn't I?" Of course the answer is "No." If they had done fewer treatments, the residual mean square and degrees of freedom which determine the LSD would have changed also!

Multiple range tests

A solution is to use a "multiple range test" in which the value of "t" is increased from that given in the standard "t" table by a compensating multiplier. This multiplier gets larger as the two means we are comparing are separated by other means between them, i.e. they are further and further apart as we rank the means in order of magnitude. The LSD then becomes:

$$\text{LSD} = t_{(P=0.05,\ \text{residual d.f.})} \times \text{compensating multiplier}$$

$$\times \sqrt{2 \times \text{residual mean square}/n}.$$

Many statistical texts contain tables of these $t \times multiplier$ numbers under index entries such as "percentage points (Q) in the studentized range" or "Duncan's multiple range test." We shall be using Table A2.2 in Appendix 2 at the back of this book. It is similar to other such tables in that the rows are applicable to different degrees of freedom, and the "number of treatments" in the different columns refers to the size of the group of means of which we are comparing the extreme members. The Table gives **$t \times multiplier$** figures for $P = 0.05$ only, other books may contain additional tables for $P = 0.01$ and $P = 0.001$.

Thus the column headed "2" refers to 2 means without any means in between, i.e. adjacent means. As the unmodified t-test is appropriate in this comparison, the numbers in the first column are identical to values for t at $P = 0.05$ in Table A2.2 in Appendix 2. Unfortunately, this is not the case in most other versions of the table (see Box 16.1). You will soon know which table you have, since the figure for $P = 0.05$ at infinite degrees of freedom for 2 means will not be the expected 1.96 found in tables of t, but 2.77!

BOX 16.1   Relation of multiple range test tables to tables of *t*

As pointed out in the text, 1.96 is the appropriate value of *t* for LSD tests between adjacent means at infinite degrees of freedom and $P = 0.05$. Yet the figure found in this position in most tables for multiple range tests is 2.77. Why?

If we divide 2.77 by 1.96, we get the answer 1.41. Table A2.2 in this book has 1.96 where many other similar tables have 2.77, If we divide any of the numbers in tables which have this 2.77 by the equivalent numbers in Table A2.2, the answer will always be 1.41.

So, what's special about this 1.41? It is the square root of the number 2. The multiple range LSD takes the form:

$$\boxed{t_{(P=0.05,\ \text{residual d.f.})} \times \text{compensating multiplier}}$$

$$\times \sqrt{2 \times \text{residual mean square}/n} \ \ \text{(page 216)}$$

where the rectangle shows what is included in the figures of Table A2.2.

The number 2 that becomes 1.41 when square rooted is there! And it is under a square root bracket. We can re-write the multiple range LSD as:

$$\boxed{t_{(P=0.05,\ \text{residual d.f.})} \times \text{compensating multiplier} \times \sqrt{2}}$$

$$\times \sqrt{\text{residual mean square}/n}$$

The rectangular frame shows the multiplier used in most other tables – it includes the $\sqrt{2}$ which is part of the normal s.e.d.m. calculation.

The s.e.d.m. ($\sqrt{2 \times \text{residual mean square}/n}$) is usually a part of a computer printout. So I find it easier to operate the multiple range test by using Table A2.2 to enlarge *t* in the standard LSD, i.e. $t_{(P=0.05,\ \text{residual d.f.})} \times$ s.e.d.m. than to go back to the residual mean square as required by most forms of the table of multipliers.

Operating the multiple range test

It is probably best to illustrate this test with some example figures, and I will describe the fullest way of using a multiple range test. This will seem pretty tedious and long-winded, and you will see later in this chapter that such a full exercise is rarely needed. However, it is worth demonstrating to you, as it will familiarize you with what steps you are leaving out when you short-circuit the technique and will help you be content that short-circuiting will not lead to false conclusions.

So here are nine imaginary means (A–I) of three replicates with a s.e.d.m. from the analysis of variance of 2.41:

A	B	C	D	E	F	G	H	I
22.00	34.80	32.00	27.30	25.10	20.00	36.20	32.40	26.50

Step 1 – re-arrange these means in order of decreasing magnitude:

G	B	H	C	D	I	E	A	F
36.20	34.80	32.40	32.20	27.30	26.50	25.10	22.00	20.00

Step 2 – repeat the treatment codes vertically to the right of the grid, starting with the largest mean:

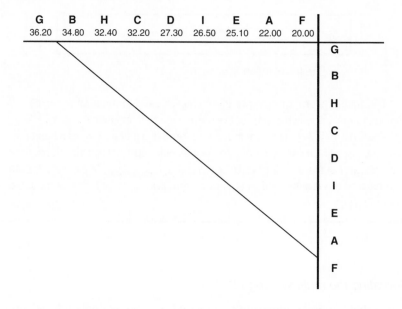

G	B	H	C	D	I	E	A	F
36.20	34.80	32.40	32.20	27.30	26.50	25.10	22.00	20.00

G
B
H
C
D
I
E
A
F

Step 3 – Fill in the code for the difference between two means represented by each position in the grid (the diagonal line shows where the same mean is at the end of both co-ordinates. Below this diagonal, differences above it would merely be repeated – e.g. B–G would repeat G–B.

Step 4 – Under each coded difference, insert the actual difference between the two means:

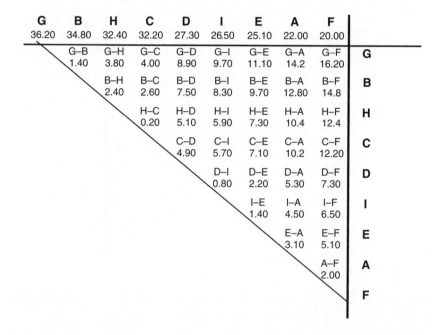

G	**B**	**H**	**C**	**D**	**I**	**E**	**A**	**F**	
36.20	34.80	32.40	32.20	27.30	26.50	25.10	22.00	20.00	
	G–B	G–H	G–C	G–D	G–I	G–E	G–A	G–F	**G**
	1.40	3.80	4.00	8.90	9.70	11.10	14.2	16.20	
		B–H	B–C	B–D	B–I	B–E	B–A	B–F	**B**
		2.40	2.60	7.50	8.30	9.70	12.80	14.8	
			H–C	H–D	H–I	H–E	H–A	H–F	**H**
			0.20	5.10	5.90	7.30	10.4	12.4	
				C–D	C–I	C–E	C–A	C–F	**C**
				4.90	5.70	7.10	10.2	12.20	
					D–I	D–E	D–A	D–F	**D**
					0.80	2.20	5.30	7.30	
						I–E	I–A	I–F	**I**
						1.40	4.50	6.50	
							E–A	E–F	**E**
							3.10	5.10	
								A–F	**A**
								2.00	
									F

Step 5 – Add dotted diagonal lines as shown in the next grid. Each diagonal identifies differences between means to which the same new "compensated LSD" will apply (i.e. the two means involved are the extremes of a group of means of the same size), beginning with adjacent means on the bottom diagonal. The numbers in brackets next to the treatment code letters in the vertical dimension identify the size of the group of means (i.e. 2 for adjacent means).

Step 6 – At each step in the vertical dimension, multiply the s.e.d.m. from the analysis of variance (2.41 in our example) by the multiplier in the table for the residual sum of squares degrees of freedom (9 treatments × 3 replicates = 8 × 2 = 16 d.f. in our example) appropriate for the right size of the group of means involved (i.e. 2 for adjacent means). Details of these calculations are in Box 16.2. Add these "new LSD" values to the right of the table:

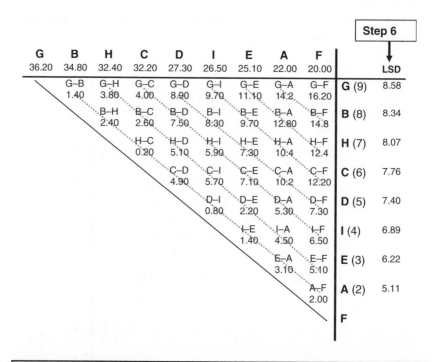

BOX 16.2   Using the table of s.e.d.m. multipliers

The worked example in the text has a s.e.d.m. from the analysis of variance of 2.41. In the multiple range test, "t" in the LSD needs enlarging by a multiplier (see Appendix A2.2) which increases as differences between means further apart in order of magnitude are being tested for statistical significance. The residual degrees of freedom are 16 (9 treatments × 3 replicates = 8 × 2 d.f.). The working to obtain these:

Size of group of means	s.e.d.m.	Multiplier from table for 16 d.f.	"New" LSD for multiple range test
2	2.41	2.12	5.11
3	2.41	2.58	6.22
4	2.41	2.86	6.89
5	2.41	3.07	7.40
6	2.41	3.22	7.76
7	2.41	3.35	8.07
8	2.41	3.46	8.34
9	2.41	3.56	8.58

(Continued)

BOX 16.2   Continued

Note that for differences between adjacent means (i.e. size of group = 2), the multiplier for the s.e.d.m. is 2.12, the same as found for 16 d.f. and $P = 0.05$ in the table of t (Table A2.1). This is because the unmodified t-test is fully appropriate for adjacent means (page 222).

This addition shows clearly the principle of the multiple range test – the least significant difference (LSD) becomes larger as the difference being tested is between means increasingly separated in the ranking by order of magnitude. Differences between adjacent means (LSD = 5.11) are in fact being tested by the unmodified t-test (see Box 16.2).

Step 7 – The differences in the table are now tested for significance against the appropriate LSD as per the dotted lines. The order of doing this is important. We begin from the right on the top line, and move horizontally left, checking the size of each difference against the LSD at the end of the relevant diagonal. We move down to the next line when the first nonsignificant difference is encountered. First, however we draw a vertical line, and do **no** more tests to the left thereof (even if a difference there appears significant). In the table below, significant differences are identified in italicized bold type:

G	B	H	C	D	I	E	A	F		LSD
36.20	34.80	32.40	32.20	27.30	26.50	25.10	22.00	20.00		
	G–B 1.40	G–H 3.80	G–C 4.00	*G–D 8.90*	*G–I 9.70*	*G–E 11.10*	*G–A 14.2*	*G–F 16.20*	**G (9)**	8.58
		B–H 2.40	B–C 2.80	*B–D 7.50*	*B–I 8.30*	*B–E 9.70*	*B–A 12.80*	*B–F 14.8*	**B (8)**	8.34
			H–C 0.20	H–D 5.10	H–I 5.90	H–E 7.30	*H–A 10.4*	*H–F 12.4*	**H (7)**	8.07
				C–D 4.90	C–I 5.70	C–E 7.10	*C–A 10.2*	*C–F 12.20*	**C (6)**	7.76
					D–I 0.80	D–E 2.20	D–A 5.30	D–F 7.30	**D (5)**	7.40
						I–E 1.40	I–A 4.50	I–F 6.50	**I (4)**	6.89
							E–A 3.10	E–F 5.10	**E (3)**	6.22
								A–F 2.00	**A (2)**	5.11
									F	

There are six nonsignificant differences (underlined) larger than 5.11, the LSD appropriate for adjacent means. Without a multiple range test we would have accepted these as valid differences. Note particularly C–E, which at 7.10 is larger than the LSD of 6.89 at the end of its diagonal – yet is declared as nonsignificant. This is because H–E, a larger difference of 7.20, has previously been declared as nonsignificant, and mean C lies between H and E!

Testing differences between means

There is no dispute about the fact that repeated chance sampling several times from one and the same normal distribution will soon produce means with differences between them that will turn up infrequently and less than once in 20 times (i.e. $P < 0.05$). Significance tests at this level of probability, such as the unmodified t-test, are therefore often inappropriate. Unfortunately, the solution provided by multiple range tests is not accepted by many statisticians, who query the validity of the assumptions made. Some statistical referees will not OK a scientific paper which uses these tests; others will not OK it if it does not!

There's little point in you or me trying to adjudicate on this matter, but I offer the following suggestions as (i) at least providing an **objective** basis for our conclusions from an experiment and (ii) erring on the side of safety. We may miss real differences, but are unlikely to claim a difference where none actually exists (see the note below). The trouble is that my suggestions may well please no one – after all the Thai proverb puts it very well: "He who takes the middle of the road is likely to be crushed by **two** rickshaws."

*Note: In any case, a lack of significance doesn't mean a difference is not real. The replication may not have been adequate to detect a difference that in truth exists (see Appendix 1), or by bad luck our sampled difference is a bad underestimate of a genuine difference. You can't **ever** use statistics to be 100% sure that two different sampled means are really estimates of an identical mean. In other words, although you can frequently be virtually sure that treatments have had an effect, you can never be as sure that they haven't!*

Suggested "rules" for testing differences between means

1 The unmodified t-test is always appropriate for testing differences between two means that are **adjacent** in order of magnitude.
2 If the means represent treatments which are progressive levels of a single factor, e.g. increasing or decreasing doses of fertilizer, a concentration

series of insecticide, etc., **and** the means follow a progression reasonably compatible with the series of treatments (e.g. plant weight seems to increase with nitrogen fertilization), then you may use the *t*-test for differences even between means that are not adjacent and so identify how far the progression has to go before the change in the means attains "statistical significance." Linear relationships between a datum and a quantitative change in a treatment are analyzed by regression or correlation (Chapter 17), but significant differences revealed by Anova are better evidence of "cause and effect" (as discussed at the start of Chapter 17).

3 If the means are of treatments of a more independent nature (i.e. they are qualitative rather than quantitative – such as different plant varieties, different chemicals) then you should consider (but see 3a and 3b below) a multiple range test for any differences between means which are not adjacent. If the experiment merely seeks to identify treatments which are superior/inferior to a single control, the unmodified LSD is often used. However, if the control is just "another treatment" in the sense that its mean has been established with no more precision than any other of the treatments, the use of the unmodified LSD test is not really appropriate.

(a) Do first check with the unmodified LSD test that tests between adjacent means do not adequately identify the significant differences that exist.

(b) If the interest in the experiment is primarily in a significant interaction(s), then it is often possible to use the unmodified LSD to pick out just the few differences (between the levels of one factor at one or two levels of the other) for which it is worth calculating the modified LSD appropriate to that test(s).

Note: The multiple range test appears pretty complicated when described as above, but is actually much quicker to do than to write about. There is also the snag that you may have to do it manually – you won't find it in many standard computer packages. What a thought – having to do some maths! But it's perhaps a good thing to move forward from the view that so many folk seem to have: about statistics and drawing graphs: "If it can't be done on a computer, then it can't be done at all."

Presentation of the results of tests of differences between means

As emphasized at the start of this chapter, a significant variance ratio for even as few as three treatments may result from only one difference between the means being significant. How do we present such results when listing the means involved?

The most generally accepted procedure is to use lower case letters "a," "b," "c," etc., whereby means which do not differ significantly share the same letter. There seems no agreed rule as to whether "a" goes to the highest or lowest mean; I tend to use "a" for the highest mean. Thus, for the case in the previous paragraph of three means (of which two are not statistically distinguishable), the listing of means might look like:

49.4   a

48.7   a

32.1   b

It is quite possible to get "overlap – for example, the highest mean might be clearly separable from the lowest, but the middle mean might not be separable from either of the others. We would show this as:

49.4   a

40.0   ab

32.1   b

To go into a bit more detail when things get more complicated, we'll take the means, and which differences between them are statistically significant, from the imaginary example I used for the multiple range test.

We first list the means in order of magnitude (I'll do this with the means across the page):

G	B	H	C	D	I	E	A	F
36.2	34.8	32.4	32.2	27.3	26.5	25.1	22.0	20.0

Beginning with the highest mean (G) we label it "a," as well as the other means which do not differ significantly:

G	B	H	C	D	I	E	A	F
36.2	34.8	32.4	32.2	27.3	26.5	25.1	22.0	20.0
a	a	a	a					

We now move to B. Only H and C are not significantly different, so the "a" label's already taken care of that. For H, we have C, D, I, and E which do not differ significantly, so it is time to introduce letter "b" as:

G	B	H	C	D	I	E	A	F
36.2	34.8	32.4	32.2	27.3	26.5	25.1	22.0	20.0
a	a	ab	ab	b	b	b		

Moving on to "C," the inseparable means are D, I, and E. H, however, is significantly different, so we show this with the distribution of "c":

G	B	H	C	D	I	E	A	F
36.2	34.8	32.4	32.2	27.3	26.5	25.1	22.0	20.0
a	a	ab	abc	abc	abc	abc		

Now we reach mean D, and find that we have come to the end of the differences that are significant, so means D–F all share the common letter "d":

G	B	H	C	D	I	E	A	F
36.2	34.8	32.4	32.2	27.3	26.5	25.1	22.0	20.0
a	a	ab	abc	abcd	abcd	abcd	d	d

Were these real results, we would point to the existence of two groups of means, the high G–C and the lower D–F, with some overlap in the middle (H–E).

The results of the experiments analyzed by analysis of variance in Chapters 11–15

This is the bit you've all been waiting for? I'm now going back to all the analyses of variance of experiments laid out in the different designs I have described, and I will in each case suggest how the differences between the means might be compared statistically and interpreted biologically.

In each case I'll first identify the size and structure of the experiment with what I called the "lead line" in earlier chapters (page 118). Then I'll extract what sources of variation were statistically significant at at least $P = 0.05$ by the F test in the analysis of variance, together with (needed to calculate the LSD) the residual mean square and its degrees of freedom.

Next I'll take each significant source of variation and calculate the appropriate LSD. To save repetition, all LSDs and claims of statistical significance (unless otherwise stated) will be at the $P = 0.05$ probability level. I'll then present, for that source of variation, the relevant mean values, usually in order of magnitude. In most cases I stopped short of calculating the treatment and block means in the earlier chapters, so here I will have had to derive from totals given in the appropriate chapter, by dividing by the number of plots involved.

I will then discuss my approach to using the unmodified *t*-test or a multiple range test to separate means statistically, and finally I will give my biological interpretation of the results. You'll notice I have used the first person a lot here – I think I have already made it clear that there is some lack of unanimity concerning this final stage of biological experiments!

But don't skip over the first example – this has more explanation than the others because it is the first; more is assumed after that.

Fully randomized design (page 117)

> **Structure:** 3 Fertilizers each occurring 4 times = 12 plots.
> **Data:** Days to 50% flowering per plot of broad bean plants.
> **From analysis table:** Fertilizer effect significant at $P < 0.01$; residual mean square of 5.50 @ 9 d.f.
> **Least significant difference (Fertilizer effect):** Means are of 4 plots. The detailed working via the s.e.d.m. will only be presented for this first example:

$$\text{s.e.d.m} = \sqrt{(2 \times 5.50)/4} = 1.658$$

residual mean square   number of plots in mean

$$\text{LSD} = 1.658 \times t_{(P=0.05,\ 9\ \text{d.f.})} = 1.658 \times 2.262 = 3.75$$

d.f. for residual mean square

Note: Whatever number of decimal places you keep during calculations, the final LSD yardstick for comparing means should have the same number of decimal places as the means being compared.

Fertilizer means in order of magnitude:

	B	A	C
	40.00	38.75	30.75
Differences between		1.25	*8.00*
adjacent means			
(bold italic type indicates			
difference > LSD)			

Interpretation of Fertilizer effect: That only the A–C difference exceeds the LSD shows that the bean plants flower significantly earlier with fertilizer C than with either fertilizers A or B, whose two means cannot be statistically separated. Letters (see page 224) can therefore be added as:

B	A	C
40.00	38.75	30.75
a	a	b

Randomized block experiment (page 123)

Structure: 3 Fertilizers × 4 Blocks = 12 plots.
Data: Days to 50% flowering per plot of broad bean plants.
From analysis table: Fertilizer and Block effects are both significant at
  $P < 0.01$; residual mean square of 8.26 @ 6 d.f.
Least significant difference (Fertilizer effect): Means are of 4 plots.

$$\text{LSD} = \sqrt{(2 \times 8.26)/4} \times 2.447 = 4.97$$

Fertilizer means in order of magnitude:

	B	A	C
	39.50	37.50	29.50
Differences between		2.50	*8.00*
adjacent means			
(bold italic type indicates			
difference > LSD)			

Interpretation of Fertilizer effect: That only the A–C difference exceeds the LSD shows that the bean plants flower significantly earlier with fertilizer C than with either fertilizers A or B, whose two means cannot be statistically separated. Letters (see page 224) can therefore be added as:

B	A	C
40.00	38.75	30.75
a	a	b

  This is the same interpretation as for the fully randomized design. Hardly surprising, since both experiments also involve the flowering of broad beans and the same three fertilizers!

Least significant difference (Block effect): These means are now of only three plots:

$$\text{LSD} = \sqrt{(2 \times 8.26)/3} \times 2.447 = 5.74$$

the only change

Block means in order of magnitude:

	4	3	2	1
	43.67	39.00	30.00	29.33

Differences between
adjacent means
(bold italic type indicates
difference > LSD)

	4.67	**9.00**	0.67

Interpretation of Block effect:

4	3	2	1
43.67	39.00	30.00	29.33
a	a	b	b

The bean plants flower significantly earlier in the pair of blocks 1 and 2 than in the pair of blocks 3 and 4. However, I note that the rank order of means follows a consistent gradient from block 1 to block 4. Therefore I would not be so bold as to claim that the difference of 4.67 between the means of blocks 3 and 4 is not a real effect, merely that the two means cannot be separated statistically at $P = 0.05$ (when the LSD is 5.74). Indeed, at $P = 0.1$ (the 10% chance) the difference of 4.67 **would** be statistically significant (see Box 16.3 – but first see if you can confirm this yourself!).

BOX 16.3

The formula in the text for the LSD at $P = 0.05$ was given as:

$$\sqrt{(2 \times 8.26)/3} \times 2.447 = 5.74$$

The key to calculating the LSD for a different level of probability lies in the number 2.447, which is the value in t tables for the six residual mean square d.f. at $P = 0.05$.

In finding the LSD at $P = 0.1$, the residual degrees of freedom are unaffected, so we need to look up the t value for the same 6 d.f. at $P = 0.1$.

We will find this to be 1.943. Therefore the LSD at $P = 0.01$ is:

$$\sqrt{(2 \times 8.26)/3} \times 1.943 = 4.56$$

Latin square design (page 132)

> **Structure:** 4 Columns × 4 Rows = 16 plots of 4 cabbage Varieties (i.e. 4 plots per variety).
> **Data:** Numbers of winged aphids arriving on 20 plants per plot.
> **From analysis table:** Variety and Rows show significant effects at $P <$ 0.001, Columns at $P < 0.01$; residual mean square of 17.42 @ 6 d.f.
> **Least significant difference:** Variety, Row and Column means are all based on the same number of plots (4), so only one LSD needs to be calculated for this experiment:

$$\text{LSD} = \sqrt{(2 \times 17.42)/4} \times 2.447 = 7.22$$

Variety means in order of magnitude:

	D	A	C	B
	85.5	51.25	39.25	35.00
Differences between adjacent means (bold italic type indicates difference > LSD)	**34.25**	**12.00**	4.25	

Interpretation of Variety effect:

D	A	C	B
85.5	51.25	39.25	35.00
a	b	c	c

The simple LSD test is all that is needed. Most aphids settled on variety D, significantly fewer on A, and least on C and B (these two adjacent means cannot be separated statistically).

Column means in order of magnitude:

Columns (1–4 from left to right)	1	2	3	4
	62.75	56.25	50.75	41.25
Differences between adjacent means (bold italic type indicates difference > LSD)	6.50	5.50	**9.50**	

Interpretation of Column effects: Only the difference between the means of columns 3 and 4 exceeds the LSD. However, there is clearly a consistent left–right gradient in the four means, and the difference of 12.00 between means 1 and 3 can also be accepted as significant, thus:

1	2	3	4
62.75	56.25	50.75	41.25
a	ab	b	c

Row means in order of magnitude:

Rows (1–4 from top to bottom)	1		2		3		4
	72.75		67.25		43.25		27.75
Differences between adjacent means (bold italic type indicates difference > LSD)		5.50		*24.00*		*15.50*	

Interpretation of Row effects: Rows 2, 3, and 4 have statistically separable and progressively decreasing means. The decrease in the mean of row 2 from row 1 is less than the LSD:

1	2	3	4
62.75	56.25	50.75	41.25
a	a	b	c

Interpretation of positional effects: The use of a Latin square design provides statistically validated evidence that row and column means for arriving winged aphids fall off in both vertical and horizontal directions from the top left-hand corner. The aphid immigration flight clearly arrived from approximately that direction.

2-Factor experiment (page 155)

Structure: 3 Fertilizers × 2 strawberry Varieties × 3 Blocks = 18 plots.
Data: Yield of strawberries (kg/plot).
From analysis table: All effects (of Fertilizer, Variety, Fertilizer × Variety interaction, and Blocks) are significant at $P < 0.001$; residual mean square of 9.0 @ 10 d.f.
Least significant difference (Variety × Fertilizer): Note that significant interactions are interpreted before overall effects of the main

factors, since the presence of the interaction can make interpretation of the overall effects misleading.

Variety × Fertilizer means are means of three plots:

$$\text{LSD} = \sqrt{(2 \times 9.0)/3} \times 2.228 = 5.46$$

Variety (Y, Z) × Fertilizer (A, B, C) means (in order of magnitude):

	BY	AY	BZ	CZ	AZ	CY
	38.67	36.67	26.33	25.00	24.67	20.00
Differences between	2.00	*10.34*	1.33	0.33	4.67	
adjacent means						
(bold italic						
type indicates						
difference > LSD)						

The only statistically valid difference between adjacent means is AY–BZ. However, BZ and CY differ by 6.33, larger than the LSD of 5.46. As these means are not adjacent but the extremes of a group of four means, a multiple range test is called for:

$$4 \text{ mean range LSD} = \sqrt{(2 \times 9.0)/3} \times 3.06 = 7.50$$

from Appendix A2.2 for 4 treatments and 10 d.f.

So BZ and CY are **not** statistically separable.

Interpretation of Fertilizer × Variety interaction (arranged in order of magnitude of means for main factors – in italic type). Re-arranging the means this way is a neat trick for making it easier to see what a significant interaction means biologically):

	B	A	C	
Y	38.67a	36.67a	20.00b	*31.80*
Z	26.33b	24.67b	25.00b	*25.53*
	32.50	*30.67*	*22.50*	

Variety Z has a uniform yield regardless of fertilizer, whereas variety Y shows yield increased over Variety Z with fertilizers A and B only, and to a similar extent. If a grower particularly wished to grow variety Z (e.g. because it fruited particularly early), he could use any of the three fertilizers. But fertilizer C must be avoided when growing variety Y! Although the main factors had a significantly high variance ratio in the analysis, it would be misleading to interpret them. The superiority of fertilizer B, suggested by the main factor means, does not apply to variety Z; similarly the suggested better yield of variety Y does not apply with fertilizer C.

Least significant difference (Block effect): The Block means are of six plots:

$$\text{LSD} = \sqrt{(2 \times 9.0)/6} \times 2.228 = 3.86$$

Block means in order of magnitude:

	3		2		1
	34.00		28.83		22.83
Differences between		**5.17**		**6.00**	
adjacent means					
(bold italic type indicates					
difference > LSD)					

Interpretation of Block effect:

	3	2	1
	34.00	28.83	22.83
	a	b	c

There is clearly a strong gradient with yields decreasing from block 3 through 2 to block 1; the means of all three blocks can be distinguished statistically.

4-Factor experiment (page 173)

Structure: 3 Varieties of leeks × 2 Fertilizers × 2 Alignments × 3 Trimmings × 3 Blocks = 108 plots.
Data: Marketable yield (kg/plot).
From analysis table: Effects of Variety, Alignment, Variety × Alignment interaction, and Blocks are all significant at $P < 0.001$; effects of Fertilizer and Trimming are significant at $P < 0.01$; no other interactions of any order reach significance; residual mean square of 0.58 @ 70 d.f.
Least significant difference (Variety × Alignment):

Variety × Alignment means are means of 18 plots:

$$\text{LSD} = \sqrt{(2 \times 0.58)/18} \times 1.994 = 0.51$$

Variety (A, B, C) × Alignment (P, Q) means (in order of magnitude):

	CP	CQ	BP	AP	AQ	BQ
	9.17	9.09	9.04	8.50	7.36	6.04
Differences between adjacent means (bold italic type indicates difference > LSD)	0.08	0.05	*0.59*	*1.14*	*1.32*	

Interpretation of Variety × Alignment interaction (arranged in order of magnitude of means for main factors – in italic type): The LSD test provides all the separation of means that is possible, without the need for a multiple range test:

	C	A	B	
P	9.17a	8.50b	9.04a	*8.90*
Q	9.09a	7.36c	6.04d	*7.50*
	9.13	*7.92*	*7.54*	

Leeks of varieties A and B, planted along the row (P), yield more than those planted with the fan across the row (Q), but variety C gives a very high and similar yield however aligned. Both alignments have lower yields when variety A is used than with C. Variety B shows a particularly strong interaction with alignment; P leeks yield as well as the top-yielding variety (C), but when Q-planted they have the worst yields. So planting along the row (P) seems the best option, but is particularly important for any grower deciding to use variety B, perhaps because of valuable disease resistance.

The overall effects of Variety and Alignment, although both have *F* values at $P < 0.001$, are too inconsistent to be interpreted separately from their interaction. Variety C does not outyield B in the P orientation, and (unlike the other varieties) is not affected by orientation.

Interpretation of Fertilizer effect: One top dressing gave a mean yield of 8.00 kg/plot, raised to 8.40 when a split dose was used. With only two means, a significantly high variance ratio in the analysis is all that is needed to confirm they are significantly different.

Least significant difference (Trimming effect): The Trimming means are of 36 plots:

$$\text{LSD} = \sqrt{(2 \times 0.58)/36} \times 1.994 = 0.36$$

Trimming means in order of magnitude:

	T		R		L
	8.43		8.31		7.86
Differences between		0.12		*0.45*	
adjacent means					
(bold italic type					
indicates difference > LSD)					

Interpretation of Trimming effect:

T	R	L
8.43	8.31	7.86
a	a	b

There is no "untrimmed" treatment, so we have to assume that all plants were trimmed because it is known that trimming gives some advantage that the grower values (even if only ease of handling at planting). The results above show that trimming just the leaves is the worst option; yield is raised when roots are trimmed (and to the same extent whether or not leaves are trimmed as well).

Least significant difference (Block effect): The Block means are of 36 plots:

$$LSD = \sqrt{(2 \times 0.58)/36} \times 1.994 = 0.36$$

Block means in order of magnitude:

	3		1		2
	57.10		47.72		42.80
Differences between		*9.38*		*4.92*	
adjacent means					
(bold italic type indicates					
difference > LSD)					

Interpretation of Block effect:

3	1	2
57.10	47.72	42.80
a	b	c

All block means differ significantly from each other. This is perhaps only to be expected, since the "blocks" were actually different sites and not, as in the other experiments, blocks set at right angles to a possible gradient on the same piece of land.

Split plot experiment (page 201)

Structure: 2 Sprayings × 3 Brussels sprout Varieties × 4 Replicates = 24 plots, with Sprayings allocated to main plots with Varieties as sub-plots.
Data: Yield of sprouts (kg/plot).
From analysis table: *Main plot analysis* – The effect of Spraying was significant at $P < 0.05$, the Replicate effect was not significant; main plot residual mean square 3.55 @ 3 d.f.

Sub-plot analysis – Varieties and the Variety × Spraying interaction were both significant at $P < 0.001$; sub-plot residual mean square of 0.48 @ 12 d.f.
Least significant difference (Variety × Spraying): Significant interactions are interpreted before overall effects of the main factors, since the presence of the interaction can make interpretation of the overall effects misleading.

Variety × Spraying means are means of four plots, and the relevant residual mean square is that in the sub-plot part of the analysis:

$$LSD = \sqrt{(2 \times 0.48)/4} \times 2.179 = 1.07$$

Variety (A, B, C) × Spraying (U, S) means (in order of magnitude):

	AS	CS	CU	BS	AU	BU
	29.70	28.70	28.65	28.43	22.95	19.18
Differences between adjacent means (bold italic type indicates difference > LSD)	1.00	0.05	0.22	*5.48*	*3.77*	

Interpretation of Variety × Spraying interaction (arranged in order of magnitude of means for main factors – in italic type): The LSD test provides nearly all the separation of means that is possible, but means AS and BS differ by 1.27, more than the LSD. The multiple range LSD for the extreme pair of four means in this case is $\sqrt{(2 \times 0.48)/4} \times 2.97$

(intersection of four treatments with 12 d.f. from Table A2.2) $= 1.45$, so the AS and BS difference is not significant.

	C	A	B	
S	28.70a	29.70a	28.43a	*28.36*
U	28.65a	22.95b	19.18c	*23.59*
	28.68	*25.45*	*23.80*	

This experiment is about controling aphids on different cabbage varieties with an insecticide. The results show that equally high yields can be obtained on unsprayed (U) as on sprayed (S) variety C; this variety is clearly tolerant or not susceptible to aphid attack. The other two varieties suffer significant yield loss if not sprayed. The main effects (though with significant variance ratios in the analysis of variance) are not consistent. Insecticide does not raise yield significantly on variety C, and the varieties do not have significantly different yields if insecticide is used.

Spare-time activities

1 Now here's something most unusual! Your challenge is first to do a factorial analysis of variance backwards, and then test differences between means as described in this chapter.

The data below were published by Saraf, C.S., and Baitha, S.P. (1982) in the *Grain Legume Bulletin*, IITA, Ibadan, Nigeria, and are the results of an experiment on the number of nodules per lentil plant with three dates of planting and five water regimes. The table gives the means of four replicates.

Planting dates were October 25 (D1), November 16 (D2), and December 7 (D3).

Watering regimes were low irrigation throughout crop growth (W1), low irrigation during vegetative growth followed by high irrigation in reproductive phase (W2), the reverse (i.e. high in vegetative and low in reproductive phase) (W3), high irrigation throughout (W4), and an unirrigated control (W5).

Planting dates	Water regimes				
	W1	W2	W3	W4	W5
D1	23.0	22.0	25.9	29.2	28.2
D2	21.6	33.4	30.5	32.2	20.7
D3	13.2	20.0	19.0	22.5	20.0

The authors also give the information that the LSD $(P = 0.05) = 5.7$.

You probably haven't any idea as to how to set about this? If you can manage without further clues, I really do take off my hat to you! But given two particular clues you might well have a go at this one. The clues can be found right at the end of Appendix 3.

By the way, there will be two sums of squares in the analysis table you will have to leave blank. These are the total and the replicate ones; neither can be calculated when it is the means of the four replicates per treatment that are the only data provided.

You are not required to interpret the results. I do give the outcome of a full 15-treatment multiple range test as part of the solution (Appendix 3), and it is difficult to find any clear pattern in how date of planting and watering regime affect nodulation.

2 Go back to the "spare-time activity" on grape varieties in Chapter 13, and use tests on differences between means to answer the following questions:
  (a) Should a grower wishing to produce a Müller Thurgau/Semillon blend use a different training system for the two grape varieties, assuming the cost of the two systems is identical?
  (b) Will a grower get a significantly different yield if he switches from traditional Huchselrebe to high-wire Pinot Blanc? If so, will it be higher or lower?

3 Go back to the "spare-time activity" with fungicidal seed dressings in Chapter 14. Interpret the significant 2-factor interaction.

4 Go back to the "spare-time activity" in Chapter 15 – the split plot experiment with potatoes. Interpret any significant effects of planting method and tuber size in the experiment.

17

Linear regression and correlation

Chapter features

Introduction

If there's one thing biologists seem to like doing, it's drawing lines through points. Instead of carrying out replicated treatments in an experimental design amenable to analysis of variance or some similar statistical procedure, they collect data and then attempt a graph with a line (usually a straight one!) to show that a change along one axis results in a consistent direction of change in the other. These are dangerous waters, but they do have a place in biological investigation, especially if we wish to use one measurement to predict another. An example of this would be when we use the color absorbance at a particular wavelength of carefully prepared standard solutions of different concentrations of a compound to calibrate a spectrophotometer. The point with the spectrophotometer example is that we know there is a cause and effect. We know that color absorbance will increase with concentration of our chemical – we are NOT using our results to prove that this is so!

Cause and effect

I'd like to contrast the spectrophotometer with the undoubted gradual decline in butterflies that can be graphed against increasing tonnage of insecticide applied on UK farms between 1945 and the late 1970s. Environmentalists have sought to use this relationship as anti-insecticide propaganda. However, the number of television licences issued per year also increased progressively over the same time period. We can assume that it is more likely to be the insecticides than the increased number of television licences, but we cannot be as dismissive of the decreased diversity and reduced abundance of food plants for the caterpillars of the butterflies caused over the same period by removal of hedges and increased use of herbicides. No, the increased use of insecticides, the increased number of TV licences, and the removal of plant diversity on farms were all correlated in time – which one is really to blame for the decline in butterflies? Or is it all of them? Or is it none of them, and something totally different like a subtle change in climate? We can only guess. It is not an experiment. We just cannot use any of these relationships to identify the culprit!

Other traps waiting for you to fall into

Extrapolating beyond the range of your data

We may be able to show a convincing statistical fit of a number of data points to a straight line, but straight lines rarely go on for ever, and we must usually be careful not to project the straight line beyond the points we have. Figure 17.1 shows an apparent relationship in a region between the population per square kilometer and the annual number of births. Extrapolating the line suggests the stupid conclusion that kilometer squares totally devoid of humans (i.e. at the zero population point on the horizontal axis) will become populated by spontaneously created children. Similarly, fitting a straight line to the reduction in mortality in successive years following the introduction of a malaria control program is bound eventually to reach zero mortality followed by, with further projection of the line, resurrection of the dead! In actual fact, such an apparent initial straight line reduction is not likely to lead to the elimination of malaria – the real line is almost certainly a curve (Fig. 17.2).

Is a straight line appropriate?

If a straight line is appropriate, then there should normally be as many points below as above it. Be suspicious if this is not so (e.g. Fig. 17.13,

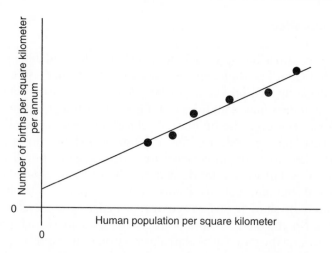

Fig. 17.1 Relationship between population density in an area and the annual number of births, illustrating the inadvisability of projecting the straight line relationship beyond the observed data points.

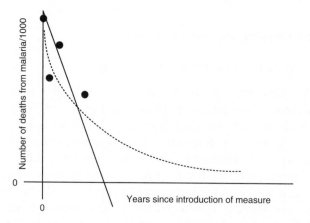

Fig. 17.2 Decline in deaths/1000 people from malaria following the introduction of a new control measure. The points in the first few years can be fitted by a straight line, but the improbability of the death rate falling to zero suggests the real relationship is a curve, or at least that the initial straight line decline will lead into a curve.

where, although a straight line has been fitted, there seems every likelihood that the true relationship is a curve and not linear!). Sometimes holding the graph paper with the points marked (before the line is drawn) horizontally at eye level, and viewing the points from the vertical axis, will show up any curved relationship quite clearly.

If a curve is indicated, then there are two possibilities. A simple hollow or convex curve without a peak can probably be converted to a straight-line relationship by transforming (page 38) the values on one or even both axes to logarithms or some other function. In testing insecticides, it is known that the frequency of insects in a population which are super-susceptible, ordinary, or resistant follows a normal distribution (page 26). This normal distribution of susceptibility translates to a sigmoid (S-shaped) increase in mortality in groups of insects treated with increasing concentrations of insecticide (Fig. 17.3a). Therefore an appropriate transformation has been established for such data. We plot what is known as the "probit" of per cent

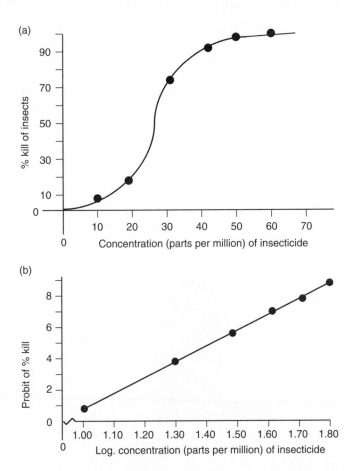

Fig. 17.3   (a) The plot of increasing per cent kill of insects against increasing concentration of insecticide. (b) By changing the scales on the axes respectively to the probit of kill ands the $\log_{10}$ concentration of insecticide, we have transformed the sigmoid curve in (a) to something close to a straight line relationship.

mortality (obtainable from statistical tables) against the logarithm to base 10 of concentration (Fig. 17.3b). This effectively straightens out the tails of the S shape of the original curve.

The advantages of a straight line compared with a curve are analogous to the advantages of the standard deviation (page 18). The straight line is defined by fewer statistics (only the slope and overall height of the line on the vertical scale – see respectively the statistics b and a later) than a curve, and has the simplest statistical recipe book to go with it.

If the curve has one or more peaks, we have to fit an appropriate relationship model with the same number of peaks (see towards the end of this chapter). You also need to be alert to the trap that data, which are clearly not related linearly, can nevertheless be fitted with a statistically validated straight line. This is because the validation is rather crude – it is that the best (however badly) fitting line is statistically distinguishable from a line with no slope (i.e. parallel to the horizontal axis). Thus Fig. 17.4 is an example from real data where a very weak negative slope can be shown to be "statistically significant" in spite of the obvious message from the data points that the point furthest to the left bucks the trend of decreasing means from left to right shown by the other three points.

Figure 17.5a, where petrol consumption of several motor cars is plotted against their length, flags another warning. At first sight a linear positive relationship seems quite reasonable. However, such a relationship assumes that the failure of the points to sit on the line is due to some error of measurement. If we look at the additional information provided for the

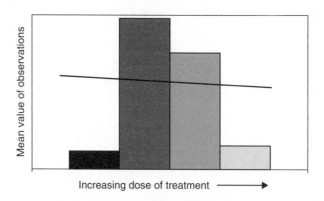

Fig. 17.4   The triumph of statistics over common sense. Each treatment mean is based on very many replicate points, and the negative relationship indicated by the straight line, between size of mean and increasing application of treatment, is actually just statistically significant!

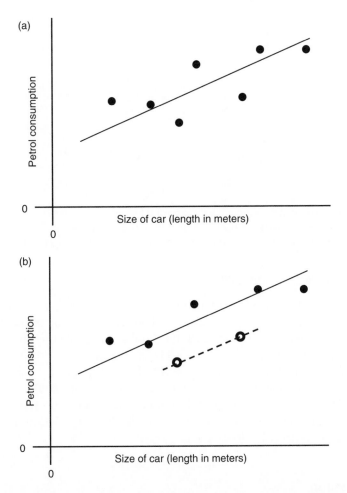

Fig. 17.5 (a) The data points would seem to warrant the conclusion that larger cars use more petrol. (b) Reveals, however, that the failure of all the points to fit exactly on the line is partly because the two open circles should not be included in the relationship – these two cars have diesel engines, which are more economical for a given vehicle size than the petrol-engined models.

same points in Fig. 17.5b, we see that the data show far more than that larger cars cost more to run. The variation about the line comes from the fact that both petrol- and diesel-fueled cars were tested, and it is clear (the two open circles) that larger diesel-powered cars are more economical than many smaller (petrol-powered) ones. The original conclusion that economy can be predicted from the length of a car is untrue, though the conclusion may be valid within either the diesel- or petrol-powered cohorts.

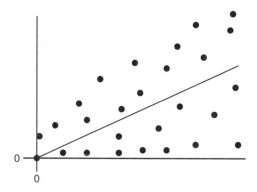

Fig. 17.6 Data which fail to meet the criterion for regression or correlation calculations that variation on the vertical scale should be equally distributed in relation to the horizontal scale. Here the variation clearly increases strikingly from left to right.

The distribution of variability

The statistics for linear relationships require that (similar as for treatments in the analysis of variance – page 103) the variability (i.e. failure of the points to sit exactly on one line) should be roughly the same across the whole graph. In Fig. 17.6 this is clearly not so: the vertical range of points increases from left to right, so that low values occur equally at all levels along the horizontal axis. However, there is the relationship that high values on the vertical axis do require high values on the horizontal axis. Transformation techniques are available for validating this latter conclusion; statistical help should be sought.

Regression

In what is called *regression*, we are trying to fit the line for which the deviations from it of the actual data points are minimized, but only in relation to one of the two axes. In the regression of y on x, we are trying to minimize the deviations from the fitted line on the vertical scale only (Fig. 17.7) – I'll explain later why we wish to do this. By convention, the vertical axis on the graph is identified as the "y" axis, with the horizontal as the "x" axis. Regression measures the slope (b) of the relationship (Fig. 17.8). As pointed out earlier, it has nothing to do with how well the points fit the line – the two top graphs of Fig. 17.8 show the same slope, but with two very different fits of the points to the line. Note also that the slope is not the slope on the graph paper – it is measured in the units of y and x. Thus both the bottom graphs of Fig. 17.8 have the same slope of $b = 0.5$, yet

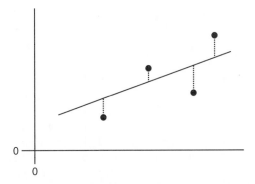

Fig. 17.7   The importance of allocating the related variables to the correct axis. The graph shows a regression of y (vertical values) on x (horizontal values) such that the regression calculation will minimize the squared distances indicated by the dotted lines. This enables the best prediction of an unknown y (the dependent variable) from a measure of x (the independent variable).

the graphs show apparently different angles for the line because the scale of the two x axes differs (look carefully!).

To measure how well the points fit the regression line, we need to calculate a different statistic from b, namely the "correlation coefficient" (r) which rises to a maximum of 1 with a perfect fit. Figure 17.8 shows lines possessing different combinations of these two statistics, b and r.

Independent and dependent variables

On a graph with two axes (x and y), each data point sits at where two measurements intersect (the y measurement and the x measurement). In *regression*, one variable is deemed to be "dependent" on the other. The latter variable (the "independent" variable) is usually graphed on the x (horizontal) axis and the "dependent" one on the y (vertical) axis. Such an arrangement is a "regression of y on x". A regression of x on y puts the variables the other way round – I'll stick with the normal type regression (y on x) in this book. If for any reason you need to do a regression of x on y (I can't actually think of a reason why you should), then merely reverse y and x in the notation that follows.

We may have measured the development time of aphids from birth to adult at different temperatures. The development time of individual aphids at any one temperature may show variation; the temperature is not varied by how long an aphid takes to develop! The temperature would be the independent variable (x axis) from which we might wish to predict a

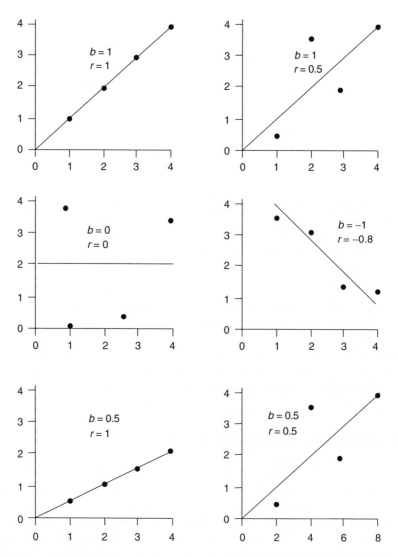

Fig. 17.8 The characteristics of the regression (*b*) and correlation (*r*) coefficients. *Top pair of graphs:* These share the same slope (*b* = 1, where an increment along the horizontal axis leads to the identical increment on the vertical axis), but the poorer fit to the points on the right reduces *r* from 1 (its maximum value) to 0.5. *Middle left:* The points have no relationship, both *b* and *r* = 0. *Middle right:* If increments on the horizontal axis are associated with a decrease on the vertical axis, both *b* and *r* have negative values. *Bottom pair:* Both graphs have the same slope (*b* = 0.5). The apparent steeper slope on the right arises because the same visual distance on the horizontal scale represents different increments in the two graphs.

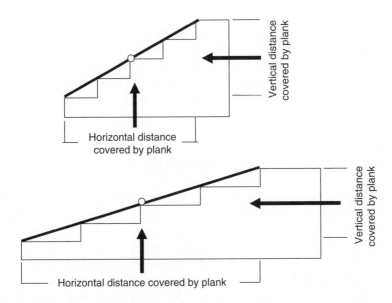

Fig. 17.9 Diagrammatic representation of a regression line as a plank laid on a staircase. The top staircase is steeper than the bottom one, but inevitably in both cases you are halfway up when you are halfway along.

development time (the dependent y variable) at a temperature intermediate between two of our experimental temperature regimes.

The regression coefficient (*b*)

The principle is nothing less familiar than a plank laid up a staircase. Figure 17.9 shows two such planks, but with different angles of slope. However, regardless of the slope, you inevitably gain half the height when you have covered half the horizontal distance. With a straight plank, that is inescapable! In terms of a graph of y (vertical axis) on x (horizontal axis), you can think of the line (the plank) as being nailed in position where halfway along the x axis (the mean of the x values) intersects with halfway up the y axis (the mean of the y values). This plank is free to swing around through 360° until we fix it in place by defining the direction of the staircase (up or down from left to right), the dimensions of the "tread" (the horizontal bit you step on), and the "riser" (the amount you rise or descend per step). With a straight line staircase (Fig. 17.9), the treads and risers are each of constant size along the flight. Thus the slope of one step fits the whole staircase. In Fig. 17.9, both planks are "nailed" at the same middle distance and height point, but the slopes of the two planks are different. This is because the top staircase has approximately equal treads and risers, and so it rises steeply. By contrast, the treads of the lower staircase are rather longer than the risers, with the result that the slope is relatively shallow. The slope of a straight line (the coefficient b) is calculated from the relative proportions to each other of the riser (units on the y axis) and the tread (units on the x axis).

Calculating the regression coefficient (b)

Slope (=the regression coefficient b) is riser/tread (see above). So if the risers are 15 cm and the treads 25 cm, the slope is $15/25 = 0.6$. Should the risers be higher (20 cm) and the treads very narrow (say only 10 cm) the slope would be $y/x = 20/10 = 2$. So in the first case, 100 cm of treads (x) would raise us by $100 \times 0.6 =$ to a y of 60 cm, and in the second case by $100 \times 2 =$ a y of 200 cm. So height can be predicted from b multiplied by the horizontal distance (x), i.e. $y = bx$. If we divide both sides of this little equation by x, it comes back to "slope = riser/tread":

$$b = y/x$$

In biology, our points rarely fall exactly on the "plank" we lay on our "staircase," and regression is the exercise of finding the "best" estimate for the slope of the plank we think the points represent. In Fig. 17.10 I show the deviations of each point (as distances in x and y units) from the one point we know sits exactly on the best fit line (i.e. as pointed out above, where mean y intersects with mean x). The figure shows clearly that we will have diverse estimates of b from each experimental point; b for the six data points varies from 0.24 to 1.17!

*Note: Read that bit again. These individual values of b (riser/tread) for each point are calculated from deviations from means. Does that phrase ring a bell? It should – sums of squares (page 21) are the sum of squares of **deviations from means**. So here's advance warning that regression calculations are going to revisit what should be familiar territory.*

How can we get our best estimate of a common slope for all these variable data? We are seeking the line from which the points give the lowest summed squared deviation measured on the y axis. We could "nail" the line on graph paper where mean y and mean x intersect, and swing it around measuring deviations as shown in Fig. 17.7 until we get the overall minimum *total*

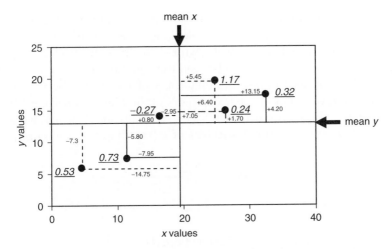

Fig. 17.10 Six data points for which one might wish to calculate a regression line. The distances of each point from the mean of their x and y values is shown as the basis for calculating b as vertical deviation/horizontal deviation. The variable b values so obtained are shown underlined.

of squared deviations from the mean of the y values (remind yourself of page 17 as to why this is the standard statistical approach to combining variation) – that would give us the best line of regression of y on x.

But of course, there has to be a computational method, and this is based on the *covariance* of the points – how the individual points co-vary on the x and y axes. This involves considering each point as based on two measurements taken from the one certain point on the final line, where mean y and mean x intersect. Thus each point has coordinates of $y - \bar{y}$ and $x - \bar{x}$ (as in Fig. 17.10).

We can't just sum all the $y - \bar{y}$ and $x - \bar{x}$ values and divide so that b (which is y/x for each point) is simply $\sum(y - \bar{y})$ divided by $\sum(x - \bar{x})$. Why not? It's because of *covariance* – any one $y - \bar{y}$ cannot be "divided" by the average $x - \bar{x}$. The contribution of that $y - \bar{y}$ to the overall slope (b) depends on its own individual distance from mean x. You may well not grasp this point at first reading, so look at Fig. 17.10. The point with a b of 0.32 only gives a b so much lower than, for example, the point with a b of 1.17 because of the difference in their respective $x - \bar{x}$ values; there is no large difference between the two points in their $y - \bar{y}$, is there?

Therefore, the computational method for calculating b has to begin by calculating the product of $y - \bar{y}$ and $x - \bar{x}$ separately for each point. Only then can these products be added together to get what is called the *sum of cross products*. In notation this is $\sum[(y - \bar{y}) \times (x - \bar{x})]$.

However, in doing this we have loused up the balance of the $b = y/x$ equation which forms the basis for calculating the slope of the regression line. By multiplying each $y - \bar{y}$ by its own $x - \bar{x}$ we have effectively multiplied the y of $b = y/x$ by x, and so changed the equation to $b = y \times x/x$! To redress this, we need to also multiply the x of the divisor by x also, i.e. $b = (y \times x)/(x \times x)$ or $b = yx/x^2$. So in notation for summed deviations from mean x, this divisor becomes $\sum(x - \bar{x})^2$. This should be immediately familiar to you as the **sum of squares of x** (see Chapter 3 if it is not!),

$$\text{so } b \text{ in notation} = \frac{\sum[y - \bar{y}] \times [x - \bar{x}])}{\sum(x - \bar{x})^2}$$

or

$$\text{in English} = \frac{\text{sum of cross products}}{\text{sum of squares for } x}$$

Let us explore this further with Fig. 17.11, which shows just eight data points in four quadrants (A–D). Each quadrant has a different combination of + and − deviations from means on the vertical and horizontal axis, so the products of these deviations will be plus where the deviations have the same sign (whether + or −), and negative where there is one + and one −. In order to keep you thinking of deviations in terms of units of x and y and not how things look on graph paper, I have introduced the nuisance factor

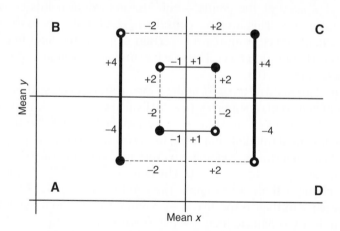

Fig. 17.11   Eight points each with a b of + or −2. Including all points in a regression calculation would show no relationship. The open circles would give a negative slope cancelled out by the positive slope of the closed ones. How the computational procedure for calculating b relates to this figure is explained in the text.

that the x and y axes are to different scales; the length on the horizontal scale that represents 2 units represents twice this (4 units) on the vertical scale.

All eight points have the same y/x of 2, though the slope of all the open circles is -2 (a slope downwards from left to right $=$ a negative slope), whereas that of all the black circles is $+2$ (a rising slope from left to right $=$ a positive slope).

The "cross product" is the two deviations from the mean y and x values for each point multiplied; thus the cross product for the top left point in quadrant B is $+4 \times -2 = -8$. For the other point in quadrat B, the cross product is -2. Two similar cross products occur in quadrat D, so the SUM of cross products for the open circles is $-8 + (-2) + (-8) + (-2) = -20$. From this *sum of cross products* we need to get to the right answer for b, which the figure was set up to equal 2 (or -2, of course). To get a slope of -2 we need to divide by 10, and this is exactly the sum of squares for x (obtained by squaring and then adding the $x - \bar{x}$ deviations for the four points $(-2)^2 + (-1)^2 + (+1)^2 + (+2)^2$).

Well, I hope the argument based on Fig. 17.11 has given you confidence that the equation:

$$\frac{\text{sum of cross products}}{\text{sum of squares for } x}$$

gives the right answer.

Let's now go back to Fig. 17.10. Does the calculation to fit the regression line really mean first working out all the "deviations" as shown on that figure and then squaring them? It's not that hard work for six points – but most regression calculations involve many more data points!

Early on in this book (page 23), we worked out a much easier way of working out the sum of squares *of deviations* from the mean of several numbers without working out the deviations first. Remember?

$$\sum (x - \bar{x})^2 \text{ could be replaced by } \sum x^2 - \frac{(\sum x)^2}{n}$$

So the divisor for the regression coefficient, $\sum (x - \bar{x})^2$, is 541.4 (see Box 17.1 for the detailed calculation).

How about the *sum of cross products*? To do this without having to work out lots and lots of individual deviations, we can modify the expression:

$$\sum x^2 - \frac{(\sum x)^2}{n}$$

BOX 17.1

Data needed for calculations of the slope (b) of the regression for the points in Fig. 17.10:

y	x	y × x
5.9	4.6	27.14
7.4	11.4	64.36
14.0	16.4	229.60
19.6	24.8	486.08
14.9	26.4	393.36
17.4	32.5	565.50
Total 79.2	116.1	1786.04
Mean 13.20	19.35	

Added squares of x:

$$\sum x^2 = 2788.33$$

Sum of squares of x:

$$\sum x^2 - \frac{(\sum x)^2}{n} = 2788.83 - \frac{(116.1)^2}{6} = 541.4$$

Sum of cross products

$$\sum ([y - \bar{y}] \times [x - \bar{x}]) = \sum (y \times x) - \frac{(\sum y \times \sum x)}{n}$$

$$= 1786.04 - \frac{(79.2 \times 116.1)}{6} = 253.5$$

which we have used throughout this book as well as in the previous paragraph, to save calculating the individual deviations of $\sum (x - \bar{x})^2$.

The thinking goes like this: $\sum (x - \bar{x})^2$ is of course $\sum ([x - \bar{x}] \times [x - \bar{x}])$, and so:

$$\sum x^2 - \frac{(\sum x)^2}{n} \text{ is also } \sum (x \times x) - \frac{(\sum x \times \sum x)}{n}$$

We can now cunningly replace one x in each part of the expression by y as:

$$\sum (y \times x) - \frac{(\sum y \times \sum x)}{n}$$

in order to calculate the sum of cross products, $\sum ([y - \bar{y}] \times [x - \bar{x}])$.

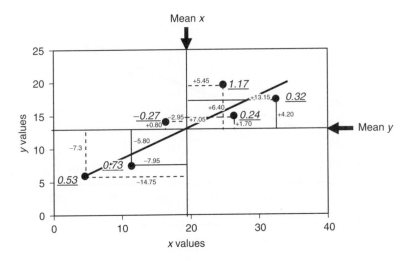

Fig. 17.12 The calculated regression line added to Fig. 17.10.

For the points in Fig. 17.10 the sum of cross products is 253.5 (see Box. 17.1), giving a slope (*b*) of $253.5/541.4 = 0.47$. Figure 17.12 adds the regression line to Fig. 17.10.

On page 23 I emphasized strongly that if you ever got a "sum of squares" that was negative, then you had made a computational error since the correction factor has to be smaller than the added squared numbers. This does NOT apply to a sum of cross products – this MUST be negative when the slope of the line is downwards (a negative slope). So don't be surprised if $(\sum y \times \sum x)/n$ is larger than $\sum(y \times x)$.

I now propose largely to give up using algebraic notation for the rest of this chapter, and propose to use (as far as possible) the following simplified terminology instead:

Sum of squares of x replaces $\sum(x - \bar{x})^2$
Sum of squares of y replaces $\sum(y - \bar{y})^2$
Sum of cross products replaces $\sum([y - \bar{y}] \times [x - \bar{x}])$.

The regression equation

It is clear that – if we were to project the slope on Fig. 17.12 to the left – it would hit the *y* axis when $x = 0$ at a value of *y* somewhat higher than 0 (at about 24 on the vertical scale in Fig. 17.13a). This value of *y* at $x = 0$ is known as the "intercept" (*a*) and is the second parameter of regression lines mentioned earlier, which determines the overall height of the line. Thus our "staircase" with slope *b* usually begins at a "landing" (*a*) above or

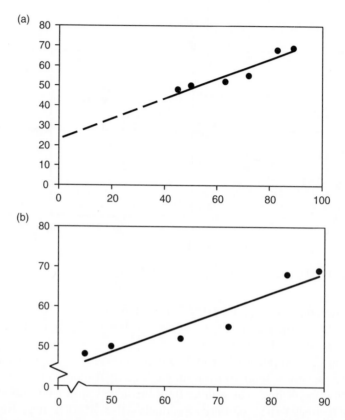

Fig. 17.13   (a) A regression line has an intercept (*a*) value at about 24 on the vertical axis at the point where the horizontal axis has its origin (i.e. where *x* = 0). (b) How graph (a) should be presented to maximize the use of the graph area (note how the fact that both axes start well above 0 can be indicated).

below *y* = 0 when *x* = 0. Thus the full regression equation for predicting *y* values from the *y* axis is:

$$y = a + bx$$

I have not yet explained how I drew the regression line in Fig. 17.12, or how to calculate the value of the intercept *a*. We also need to know how to check whether there is a statistically significant relationship between *y* and *x*, or whether our estimate of *b* is just a bad estimate of no slope (i.e. a *b* of zero). Then we also want to work out the correlation coefficient (*r*) to see how closely the points fit our line. I'll take you through all this with another set of data, this time a real set.

A worked example on some real data

The data (Box 17.2)

In a pot experiment with many replications of different densities of mustard seeds, it was suggested by someone that it would be too tedious to count out the seeds individually, and that "guesses" might be near enough. One person therefore took samples of mustard seeds from a bowl

---

BOX 17.2

Data (in bold) comparing the actual number of mustard seeds in a teaspoon (y) with the number guessed when the sample was taken.

	Actual (y)	Guess (x)	$y \times x$
	8	8	64
	15	10	150
	10	14	140
	19	18	342
	24	20	480
	20	23	460
	28	26	728
	36	30	1050
	32	36	1152
	45	43	1935
	47	50	2350
Total	283	278	8851
Mean	25.73	25.27	
n	11	11	11

Added squares of x:

$$\sum x^2 = 8834$$

Sum of squares of x:

$$\sum x^2 - \frac{(\sum x)^2}{n} = 8834 - \frac{(278)^2}{11} = 1698.82$$

Sum of cross products:

$$\sum (y \times x) - \left(\sum y \times \sum x\right)/n = 8851 - (283 \times 278)/11 = 1698.82$$

with a teaspoon, made a quick guess of the number of seeds in the spoon, and then passed the spoon to a colleague for an accurate count of the seeds it contained. A regression analysis of the actual number of seeds (y) on the guess (x) was carried out to see how accurately the actual number could be predicted from the instant guess. The data are shown in bold in Box 17.2, which also has (not in bold) some further calculations.

Totalling the y and x columns shows that guessed 278 seeds compares well with the actual 283!

Calculating the regression coefficient (b) – i.e. the slope of the regression line

The two numbers we need for this are the sum of *squares*(of deviations from the means remember) *of the guessed (x) values* and the *sum of cross products*. I am sure you remember that such calculations are most easily done by adding numbers first before subtracting a "correction factor," but in any case it really is worth following the steps by continual back-reference to the earlier section on "**calculating the regression coefficient**" in this chapter (page 248).

For the *sum of squares for x*, the added numbers are the squared x values. ($\sum x^2$) is $8^2 + 10^2 + 14^2 + \cdots + 50^2 = 8834$, and for the *sum of cross products* the added numbers are each y multiplied by its x (i.e. $x \times y$, a column added on the right of Box 17.2). The sum of these $y \times x$ values is $64 + 150 + 140 + \cdots + 2350 = 8851$.

Now for the two correction factors. The correction factor for the *sum of squares for x* is the total of x values squared and divided by n, i.e. $(278 \times 278)/11 = 7025.82$. For the *sum of cross products* the correction factor is the total of the y values multiplied by the total of the x values, and then this product is divided by n, i.e. $(283 \times 278)/11 = 7152.18$.

So $\sum(x - \bar{x})^2$ works out at $8834 - 7025.82 = 1808.18$ and $\sum([y - \bar{y}] \times [x - \bar{x}])$ at $8851 - 7152.18 = 1698.82$. Therefore b is:

$$\frac{\text{sum of cross products}}{\text{sum of squares for } x} = \frac{8851 - 7152.18}{8834 - 7025.82} = \frac{1698.28}{1808.18} = \mathbf{0.94}$$

In other words, each extra seed in the guess is on average worth just less than one actual seed – very close indeed!

Calculating the intercept (a)

As pointed out earlier, it is rarely the case that $y = 0$ when $x = 0$, even when we would expect it biologically. As explained there, the linear relationship we can show with our data may actually be part of a curve or may change from linear to a curve at some point not covered by our data. Sampling errors can also lead to us not getting the regression equation quite right – with a small error showing as a small but nonsensical y value at $x = 0$. At the start of this chapter I gave the example of zero parents nevertheless having some children, and if you complete the spare-time activity at the end of this chapter, you'll discover that a bean can weight **less** than nothing when it has zero length! So we cannot assume that our linear regression will accurately predict no seeds in an empty teaspoon. Of course, the intercept a can also be a negative number, i.e. less than no children when there are no parents! This is a cogent argument for not projecting the regression line beyond the range of our data, and for obtaining the intercept a by calculation rather than graphically.

We have already established that the regression equation is:

$$y = a + bx,$$

and we now know b to be 0.94. So $y = a + 0.94x$ for our mustard seeds.

The equation $y = a + bx$ can be re-written by subtracting bx from both sides to solve for a, i.e. $y - bx = a$. Swapping the two sides of the equation over, and inserting our calculated slope of 0.94 in place of b, we get $a = y - 0.94x$. If we had numbers for y and x, a would remain the one "unknown" and therefore can be discovered. Well we **do** have one data point where we know the y and x values without knowing a. Remember, all regression lines pass through where the means of y (=25.73) and x (=25.27) intersect, where we "nail" our plank before we swing it around to find the best fit to the points. So we can insert these two mean values into the equation $a = y - 0.94x$ above to solve for a, i.e.:

$$a = \overset{\bar{y}}{25.73} - (\overset{b}{0.94} \times \overset{\bar{x}}{25.27}) = 25.73 - 23.75 = 1.98$$

So there are nearly two seeds in an empty teaspoon!

Drawing the regression line

Now we have two points on the regression line. One is where mean y and mean x intersect, and the other is a where $x = 0$. Now we can surely draw our straight line, can't we? Well, actually that's not a very good idea, for two reasons:

1 Three points are better than two – they will show up if we have made an error in our calculations by not being exactly in a straight line, whereas you can always draw an exactly straight line between two points, even if one of them is wrong!
2 Often our data points are nowhere near $x = 0$, and including the intersection of $y = 0$ and $x = 0$ would leave most of the graph area unproductively blank – just compare Figs 17.13a and b in terms of use of graph space and usefulness of the presentation.

A better approach is therefore to retain the intersection of mean y (25.73) with mean x (25.27) as one of our points, but to calculate two further points by substituting two values for x (anywhere within the range of x but well apart on either side of the mean) in the regression equation. Now we know a to be 1.98, the new y point of the regression line will be the one "unknown" we can solve for. Given that the minimum and maximum x values for the mustard seeds are 8 and 50, we could use say 10 and 50, i.e.:

$$\overset{a}{}\quad\overset{b}{}\quad\overset{x}{}$$
$$y = 1.98 + (0.94 \times 10) = 11.38 \quad \text{and}$$
$$y = 1.98 + (0.94 \times 50) = 48.98$$

With the three intersections of $y = 25.73$ at $x = 25.27$, $y = 11.38$ at $x = 10$, and $y = 48.98$ at $x = 50$, we can plot the points and draw in the accurate line for predicting the actual number of mustard seeds (y) corresponding to a guess (x). The result is Fig. 17.14.

Testing the significance of the slope (b) of the regression

The regression calculations will almost always give a $+$ or $-$ number for b, even if there is no real evidence of a relationship (i.e. the slope is not statistically distinguishable from a zero slope). In Fig. 17.11, data points in quadrants A $+$ C (the closed circles) or B $+$ D (the open circles) have respectively strong positive and negative slopes, but the points in A $+$ B and C $+$ D or A $+$ D and B $+$ C have zero slope. Any line we drew would be parallel to either the horizontal or vertical axis, and therefore would have no slope.

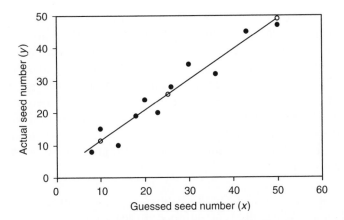

Fig. 17.14   The calculated regression line for the data points relating the actual number of mustard seeds in samples with the number simply guessed. The open circles on the regression line are the three calculated points used to draw the line.

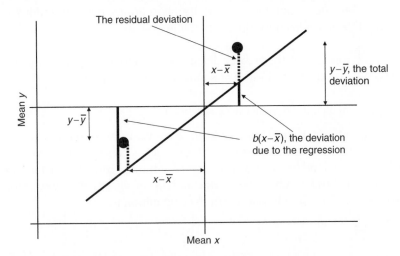

Fig. 17.15   The distances for which sums of squares are calculated in the regression analysis of variance (see text).

To test whether the regression has a significant slope, we can use analysis of variance. Look at Fig. 17.15, which identifies three sources of deviation:

1 The deviation of any point on the y axis from the mean of y (*on the figure, this distance is shown by the thin line identified as $y - \bar{y}$ running from mean y to the data point*) can be divided into the other two sources of deviation.

2 The first of these other two sources of deviation is the difference between the mean of y and where the point "should be" on the regression line at the same x value, i.e. at $b \times (x - \bar{x})$ *(on the figure, this distance is shown by the thick line identified as $b(x - \bar{x})$ running from mean y to the regression line)*. This is the part of the variation accounted for by the regression.

3 The other source is the difference between the point where it "should be" on the regression line and where it actually is *(on the figure, this distance is shown by the dotted line running from the regression line to the data point)*. This is the residual variation for that point. So sources 2 and 3 add up to 1.

For each point, therefore, we have the three components (1), (2), and (3) needed for a simple analysis of variance of the form:

Source of variation	Distance on Fig. 17.15
(2) Fit to regression line	thick line $b(x - \bar{x})$
(3) Deviation from line	dotted line (found by subtraction)
(1) Deviation from mean y	thin line $y - \bar{y}$

So now, for each of these three distances, we have to calculate, using all the points, the appropriate sum of squares *of these deviations*.

The total sum of squares for deviations from mean y

In calculating the regression coefficient b above, we were reminded about the quick way we use to sum *squared deviations* of a set of numbers from their mean – it is *added squares of the numbers* minus *a correction factor* based on the total of the numbers. So for the *total sum of squares of deviations* from mean y (i.e. $\sum y - \bar{y})^2$ in notation), the calculation is:

$$\sum y^2 - \frac{(\sum y)^2}{n}$$

Using the y column in Box 17.2, $8^2 + 15^2 + 10^2 + \cdots + 47^2 = 8893$ and the correction factor is $283^2/11 = 7280.82$. So the total sum of squares that regression and deviation sums of squares will add up to in our analysis of variance is $8893 - 7280.82 = 1612.18$, thus:

Source of variation	Sum of squares
Fit to regression line	(to be calculated)
Deviations from line (=residual variation)	(found by subtraction)
Deviation from mean y	1612.18

The sum of squares for the regression line

Look again at Fig. 17.15 and take just the right-hand point. For any data point, the x value will have moved from mean x by the amount $x - \bar{x}$, and the slope of the regression (b) will have caused y to increase from mean y by $b \times$ that x distance, i.e. $b(x - \bar{x})$. So the sum of squares for all these distances is $\sum[b(x - \bar{x})]^2$. Well, to calculate b, we already had to work out the *sum of squares of x*, as 1808.18. As this is $\sum(x - \bar{x})^2$, multiplying it by b^2 will get us to the $\sum[b(x - \bar{x})]^2$ that we're after! Simple! As b was 0.94, our sums of squares for the regression line are $0.94^2 \times 1808.18 = 1597.70$.

*Note: Statistics books usually give a different expression for the sum of squares of the regression line. They give it as the **square** of the sum of cross products divided by the sum of squares of x. This is exactly the same as $\sum[b(x - \bar{x})]^2$, see Box 17.3.*

BOX 17.3

The identity of $\sum[b(x - \bar{x})]^2$ and squared sum of cross products divided by the sum of squares of x.

The expression often used in statistics texts for the regression slope b is in notation:

$$\frac{\sum([y - \bar{y}] \times [x - \bar{x}])}{(x - \bar{x})^2} \quad \text{which we can also write as} \quad \frac{sum\ of\ cross\ products}{sum\ of\ squares\ of\ x}$$

The sum of squares accounted for by the regression, $\sum[b(x - \bar{x})]^2$, can also be written (see above) as $b^2 \times$ *sum of squares of x*.

As a square is a number multiplied by itself, we can also write this last expression as:

$$\frac{sum\ of\ cross\ products}{sum\ of\ squares\ of\ x} \times \frac{sum\ of\ cross\ products}{sum\ of\ squares\ of\ x} \times sum\ of\ squares\ of\ x$$

We can now cancel out the far right *sum of squares of x* with one of the divisors to leave:

$$\frac{sum\ of\ cross\ products \times sum\ of\ cross\ products}{sum\ of\ squares\ of\ x}$$

which is the **square** of the *sum of cross products* divided by the *sum of squares of x*.

*Completing the analysis of variance to test the significance of the
slope b*

As the residual variation is found by subtraction, we can now complete the
analysis of variance table, including the End Phase (page 99):

Source of variation	d.f.	Degrees of freedom	Sum of squares	Mean square	Variance ratio	P
Regression (slope *b*)	1	1597.70	1597.70	992.36		<0.001
Residual	9	14.48	1.61			
Total	10	1612.18				

Degrees of freedom: 11 data points give 10 d.f. for the total sum of
squares. All of this is residual (i.e. 9 d.f.) except for the single degree
of freedom represented by *b*, the regression slope we are testing for
significance.

Clearly we have a hugely significant variance ratio. The linear regression
therefore has a significant slope – there is a strong relationship between the
actual number of seeds in the spoon and our guess.

How well do the points fit the line? – the coefficient of determination (r^2)

As mentioned near the start of this chapter, the correlation coefficient, and
not the regression coefficient, is the statistic which measures how well the
points lie on our straight regression line rather than whether or not the line
has a statistically significant slope.

If there is perfect correlation, i.e. all the points sit exactly on the line,
then in our analysis of variance above all the variation in *y* values from
mean *y* are accounted for by the line. In terms of Fig. 17.15, the thick
and thin lines are identical for each point – there will be no dotted lines
(i.e. no residual variation). As points increasingly deviate from the line, the
residual will increase from zero at the expense of the sum of squares for
regression. Therefore the proportion of the total variation accounted for
by the regression line measures "goodness of fit" – this proportion will be
100% if there is no residual variation. In our mustard seed example the
total variation of 1612.18 accounted for by the regression (1597.70) is
99.1% – very high; the points fit the line very well indeed (Fig. 17.14).
99.1% can also be expressed as 0.991 out of a maximum of 1, and this
value of 0.991 (in our example) is called the *coefficient of determination* or
r^2 in notation.

Correlation

As this coefficient measures how well the points fit the line, there is no real distinction between deviations from it in both x and y directions. So whereas the regression coefficient answers the question "HOW are x and y related?," *correlation* answers a different question – "How WELL are x and y related." There are some biological situations where the appropriate analysis of relationships is with correlation rather than regression, because neither of the variables x and y are dependent on the other either biologically or for purposes of prediction. We are then not assigning dependence and independence to the two variables, definitions that are implicit in regression analysis.

Perhaps the easiest way to be sure as to whether correlation or regression analysis is appropriate is to ask the question "Do I want to know how much one changes as the other changes?" If the answer is "Yes," regression is indicated.

We measure the degree of *correlation* by the *correlation coefficient* (r). Well, following our regression analysis above, we calculated the goodness of fit of the points to the regression line by the *coefficient of determination* (r^2). Is r the square root of r^2? Is the *correlation coefficient* the square root of the *coefficient of determination?* The answer is yes, and either can simply be calculated from the other.

Derivation of the correlation coefficient (r)

In doing our analysis of variance for a regression of y on x, it became important whether *the sum of squares of x* or y was used as the divisor for calculating b, and whether the *total sum of squares* to be explained by the analysis was the sum of squares of x or y.

Thus the coefficient of determination for the regression of y on x was calculated as:

$$\frac{sum\ of\ cross\ products^2}{sum\ of\ squares\ of\ x} \quad divided\ by\ sum\ of\ squares\ of\ y$$

This relatively simple expression for the coefficient of determination may come as something of a surprise, given that we got r^2 from the analysis of variance by working out what proportion (that's the "divided by" bit) of the *sum of squares of y* was represented by $\sum(b^2 \times sum\ of\ squares\ of\ x)$. So see Box 17.3 to confirm the *sum of cross products*^{2} divided by the *sum of squares of x* bit.

Had we been working with the regression of x on y (rather than y on x), the *coefficient of determination* (r^2) would have been:

$$\frac{sum\ of\ cross\ products^2}{sum\ of\ squares\ of\ y} \quad \text{divided by } sum\ of\ squares\ of\ x$$

The *coefficients of determination* of both regressions turn out to be identical, since both can be re-written as:

$$\frac{sum\ of\ cross\ products^2}{sum\ of\ squares\ of\ x \times sum\ of\ squares\ for\ y}$$

So our *correlation coefficient (r)*, the square root of the *coefficient of determination* (r^2), has the property we expect (see first sentence of this section) that it involves no assumptions about whether y regresses on x or x on y.

r can therefore be calculated directly from data, without a preceding regression analysis, from the above formula, which can also be "square-rooted" to read:

$$\frac{sum\ of\ cross\ products}{\sqrt{sum\ of\ squares\ of\ x \times sum\ of\ squares\ for\ y}}$$

The maximum value of r is 1 (like the coefficient of determination), representing 100% fit of the points to the line, but like the regression coefficient (though unlike the coefficient of determination, which is a squared value) it can be positive or negative depending on the sign of the sum of cross products. This has to be checked when deriving r from any expression using the square of the sum of cross products, as this will be positive whether the un-squared sum is positive or negative.

The statistical significance of r can be checked in statistical tables for $n - 2$ d.f., i.e. 2 less than the number of data points. Just as with regression, a straight line accounts for 1 d.f. As d.f. for all the data points is 1 less than the number of points, $n - 2$ is left for the residual variation from the line.

An example of correlation

We might wish to test how far there is a relationship between the length and breadth of leaves of the leaves on a cultivar of *Azalea* based on measuring 12 leaves. There is no reason why one dimension should be designated the variable which is dependent on the other independent one; they are pairs of characteristics "of equal status" from individual leaves. The data and some of the calculations we require are shown in Box 17.4. From

---

BOX 17.4

Data of length and width (both in cm) of Azalea leaves.

	Length (y)	Width (x)	$y \times x$
	11.0	3.8	41.80
	11.7	5.3	62.01
	12.1	4.6	55.66
	8.0	4.2	33.60
	4.1	1.6	6.56
	7.0	2.8	19.60
	8.4	3.3	27.72
	6.8	3.0	20.40
	9.5	3.7	35.15
	6.2	3.2	19.84
	7.4	3.1	22.94
	10.9	4.0	43.66
Total	103.1	42.6	388.88
Mean	8.59	3.55	

Added squares

$$\sum y^2 = 953.17 \qquad \sum x^2 = 161.16$$

Sum of squares

$$\sum y^2 - \frac{(\sum y)^2}{n} = 953.17 - \frac{(103.1)^2}{12} = 67.37$$

$$\sum x^2 - \frac{(\sum x)^2}{n} = 161.16 - \frac{(42.6)^2}{12} = 9.93$$

Sum of cross products

$$\sum (y \times x) - \frac{(\sum y \times \sum x)}{n} = 388.88 - \frac{(103.1 \times 42.6)}{12} = 22.88$$

these calculations:

$$r = \frac{sum\ of\ cross\ products}{\sqrt{sum\ of\ squares\ of\ x \times sum\ of\ squares\ for\ y}}$$

$$= \frac{22.88}{\sqrt{9.93 \times 67.37}} = 0.885$$

This represents a high degree of correlation between leaf length and breadth, and this is also shown by the plot of the points in Fig. 17.16a, which shows as the points forming a clear ellipse.

Is there a correlation line?

If r is 0.885, then surely we can square it to get to an r^2 of 0.783, and r^2 is the *coefficient of determination* in regression – from which we can deduce that the line which the points fit accounts for 78.3% of the variation of the points from their mean. In regression, the mean in question is that of the dependent variable (y in a regression of y on x).

But what is the correlation line, and which variable does the mean relating to the r^2 refer to? There are always two lines (Figure 17.16b), depending on which way round we hold the paper (compare Fig. 17.16b and c), and so which axis forms the horizontal treads of our "staircase" (page 247). These two lines are of course the lines for the regressions of y on x and x on y. The only time there is only one line is when there is perfect correlation, when the two regressions fall along the same line. Otherwise, although both regression lines pass through both mean x and mean y, their slope differs increasingly as the level of correlation deteriorates. The calculations for both regressions in our *Azalea* leaf example are given in Box 17.5, and you will see that – although the slopes (b) and intercepts (a) differ considerably (b is 2.304 for y on x and much smaller at 0.340 for x on y; a is respectively 0.413 and 0.629) – the *coefficient of determination* (r^2) for the two regressions is identical at 0.782.

To avoid confusion, I suggest that you calculate the *coefficient of determination* (r^2) only for regressions, and the *correlation coefficient* (r) only when quantifying the association between variables which have no dependence on each other. This is not to say that you should not carry out regression analysis with such variables for "prediction" purposes. In the *Azalea* leaf example, you might wish to measure only the width and use this to convert to length – I can't think why, but you might if an insect had eaten chunks out of some leaves, making measurement of many of the widths difficult?

Extensions of regression analysis

There are some very useful extensions of regression, which are beyond the scope of this elementary text, but it is important that you should know about them. After having worked through this book, you should have the confidence to be able to use at least some of the relevant "recipes" given in

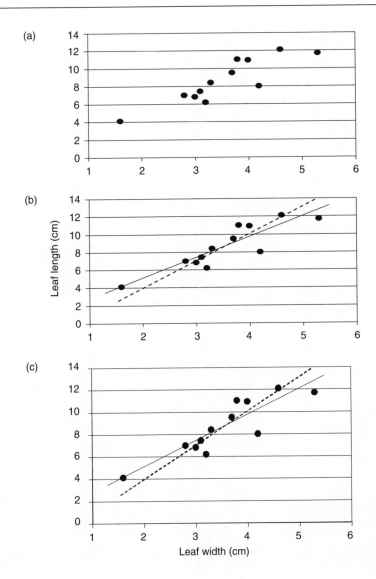

Fig. 17.16 (a) The correlation between length (y) and width (x) of *Azalea* leaves. (b) Regression lines fitted to these data; the solid line shows regression of y on x and the broken line the regression of x on y. (c) Graph (b) re-oriented to make y the horizontal axis.

BOX 17.5

Regressions using *Azalea* leaf data from Box 17.4, and using the statistics already calculated there. Dv = dependent variable in each case; Iv = independent variable.

Regression	Length (*y*) on width (*x*)	Width (*x*) on length (*y*)
Regression coefficient (*b*):		
$\dfrac{\text{Sum of cross products}}{\text{Sum of squares for Iv}}$	$\dfrac{22.88}{9.93} = 2.304$	$\dfrac{22.88}{67.37} = 0.340$
Intercept (*a*):		
Mean Dv $- (b \times$ *mean* Iv)	$8.592 - (2.304 \times 3.550) = 0.413$	$3.550 - (0.340 \times 8.592) = 0.629$
Regression equation:		
Dv $= a + (b \times$ *mean* Iv)	$y = 0.413 + (2.304 \times x)$	$x = 0.629 + (0.340 \times y)$

Regression analysis table:

Source of variation	d.f.	Sum of squares	Mean square	*F*	Sum of squares	Mean square	*F*
Regression ($b^2 \times$ *sum of squares for Iv*	1	52.71 ($2.30^2 \times$ 9.93)	52.71	35.99	7.79 ($0.340^2 \times$ 67.37)	7.79	35.41
Residual (*by subtraction*)	10	14.66	1.47		2.15	0.22	
Total (*sum of squares for* Dv)	11	67.37			9.93		
Coefficient of determination (r^2)*		$\dfrac{2.88^2}{67.37 \times 9.93} = 0.782$			$\dfrac{2.88^2}{9.33 \times 67.37} = 0.782$		

* $\dfrac{\text{sum of cross products}}{\text{sum of squares for Dv} \times \text{sum of squares for Iv}}$

larger textbooks – alternatively you will at least have realized that these are statistical techniques you might wish to talk to a statistician about.

Nonlinear regression

This takes us into the realms of nonlinear statistics. There is still only one intercept in the nonlinear regression equation, and only one y to be predicted, but x appears (raised to different powers) in the equation as many times as define the type of curve. Each time x has a different regression coefficient (b).

Thus the simplest curve, the "second order," would have a regression equation of the form $y = a + b_1 x + b_2 x^2$. Such a curve is shown in Fig. 17.17, but note that values of b can be mixed positive and negative ones. A "third order" curve has a regression equation of the type $y = a + b_1 x + b_2 x^2 + b_3 x^3$. I can probably stop with a fourth order curve at $y = a + b_1 x + b_2 x^2 + b_3 x^3 + b_4 x^4$? You'll have got the idea, I'm sure. Examples of a third and of a fourth order curve are also shown in Fig. 17.17.

The calculations for working out the regression coefficients (b) for each x term of the equation, and the analysis of variance apportioning sums of squares to each element in the regression equation are fearsome challenges

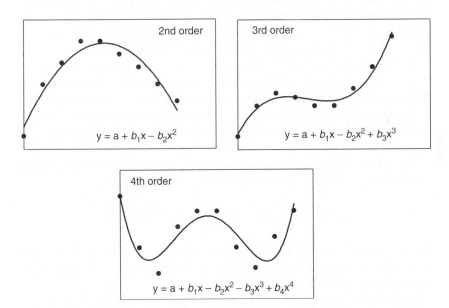

Fig. 17.17 Examples of nonlinear regression lines together with the form of their regression equations.

if done manually, and venturing into nonlinear statistics before the advent of suitable computer programs was not something to be done lightly! It was made rather easier if the x values for which y had been measured were at equal increments along the x axis. Then one could use tables of **orthogonal polynomials**, which made it relatively quick to do the calculations on a pocket calculator. The tables provide integer multipliers for each x value at each component of the equation and, from these new x values, *sums of cross products* and *sums of squares of* x for each b can be calculated. However, even then, further steps are needed before final values for the regression coefficients can be obtained.

Fortunately, computer programs have made the tedious manual computations redundant, though it is still worth bearing in mind that it pays to measure y at equal increments of x whenever this is compatible with the aims of the data collection.

Multiple linear regression

You need to know this exists – it is a powerful and very useful technique. It enables the relative importance of each of a number of independent variables (x) in determining the magnitude of a single dependent variable (y) to be estimated. I have used this wording because it does make the potential of the technique clear although the wording is bad in that it suggests "cause and effect," which of course can never be proved from the existence of a numerical relationship (as stressed at the start of this chapter).

As an example, I can cite my own interest in regressing the weight increase of greenfly against the concentrations of 14 different amino acids in plant leaves – how many amino acids, and which, have the closest positive or negative relationship with aphid weight gain? Some might be essential nutrients, others might stimulate the aphid to feed, and yet others might be deterrent or toxic.

In multiple regression we therefore have any number of independent variables, x_1, x_2, x_3, etc. So the multiple regression equation:

$$y = a + b_1x_1 + b_2x_2 + b_3x_3 + b_4x_4 + b_5x_5 \text{ etc.}$$

at first sight looks very similar to a nonlinear equation, but the numbers for x are now subscripts (to indicate different variables) and not superscripts (different powers of the same variable).

Again, before computer programs, the calculation of multiple regressions was a lengthy and complex process. It involved setting up matrices and then inverting them to solve a series of simultaneous equations; three or four independent variables were as many as most of us would have the courage to tackle. I'm told that one statistician claimed to be able to do only seven in a week!

Computer programs have made it all so easy, and there is no reason for not incorporating as many independent variables as you wish. My 14 amino acids is, by today's standards, quite a modest number!

However, there are two things you need to beware of:

Firstly (and this also applies to nonlinear regression), you need to be sensible about how many independent variables (or coefficients) you include in your final prediction of y, especially if there is only one unreplicated y measurement going into the total sum of squares (=*sum of squares of y*). You are then bound to get a perfect fit if you use all of them. Having used up all the total degrees of freedom as regression d.f., all the variation in y (total sum of squares) will have been apportioned somewhere among the independent variables! Not only do you then have no residual mean square to test the significance of the regression (since the total sum of squares = the regression sum of squares) but the fitted equation will be tied to past data and be pretty useless for predicting into the future. The literature is full of examples where workers have fitted too many variables (e.g. weather data) in their attempt to match fluctuations in the abundance of a particular crop pest over many years. They have found their "model" matches the pest's population fluctuations remarkable well – but they have then failed totally to predict fluctuations that then occurred in the next few years.

Secondly, there is always internal correlation between the variables. A plant rich in one amino acid (in my greenfly example) may well be rich in amino acids in general; hot years are likely to be drier, etc. The variable with the greater range in values will "capture" sums of squares from the other, even though the latter may be biologically more important in spite of smaller fluctuations.

Using a "stepwise" multiple regression computer program is therefore highly recommended as a safeguard against both these problems. A "stepwise" program begins by regressing y on all x variables independently. It then regresses y on all permutations of two variables. If no pair of variables improves the proportion of y variation (as sums of squares) accounted for by the regression by more than a fixed amount (often set at 5%), the program goes no further. If pairs do improve the proportion of y variation accounted for to a sufficient extent, the program next compares all permutations of 3 x variables with the results of the best paired regression – and so on. When the program stops (since it cannot further improve the regression sum of squares by the set target), you get a printout of the regression equation so far.

A safeguard against the picking of spurious independent variables, which have merely captured sums of squares from others, is to repeat the stepwise process in the opposite direction, and then focussing on those x variables which appear in both printouts. The "opposite direction" means beginning with a multiple regression on all independent variables, then trying all permutations of leaving out one variable at a time, and only going on to all permutations of leaving out two variables at a time if no regression sum of squares with one variable omitted loses more than the target (say 5%) proportion of the total sum of squares for y.

Multiple nonlinear regression

I guess computer programs make this possible, but I still think my advice should be "don't even go there"!

Analysis of covariance

This is an exceptionally valuable combination of regression analysis with analysis of variance. Sometimes the variability of your experimental material makes you plan for an analysis of covariance from the outset. However, even if you don't plan to use it, knowing it exists will encourage you to take extra data from your experiments which might prove useful should an analysis of covariance look advisable later.

What analysis of covariance is, and when one might use it, is perhaps best explained by two examples.

We might want to test the relative success of two methods of teaching French to adults – one is traditional teaching of grammar and vocabulary largely in English, and the other just speaking with them in French with no English explanations. We would be naïve not to be aware that the adults we propose to divide at random into two groups may have learnt French to different levels in school, and that many will anyway have been to France on holiday. Such variation in previous experience of the language will add undesirably to the variation of the results of a French test we plant to analyze at the end of the experiment; we say such variation is a "confounding" variable. So it makes sense to give our adults a test BEFORE we start teaching them, and use the results of this preliminary test as a *covariate* in the final statistical analysis. Putting it simply, analysis of covariance allows us to make a "correction" of the final test results for each individual, so that the results become closer to what each person would have achieved had they started with the same previous experience.

As a second example, we might plan to do a perfectly normal plot experiment on the effect of different levels of weed control on the yield of potatoes. During the experiment, some plots look stunted, and testing the soil shows that a pest, the potato cyst eelworm, patchily infests the field. This patchy infestation is clearly going to introduce serious undesirable variability into our results. Knowing of *analysis of covariance* should motivate us to measure eelworm infestation in each of our experimental plots. We can then use this as a *covariate* in the analysis of variance to correct the yield of each plot nearer to what it would have yielded if uninfested by eelworm.

The underlying concept of the analysis of covariance can be described with reference to Fig. 17.15. The covariation between the experimental data as the dependent variable (y) and the extra measurement made (x) is used to determine how much of the total variation (the sum of squares of the distances represented on Fig. 17.15 for any one point as $y - \bar{y}$) is due to regression (based on the distances labelled $b(x - \bar{x})$. This leaves the distances marked as "the residual derivation" to be partitioned into treatment and residual sum of squares as in a traditional analysis of variance. The diagram below shows what is happening:

Source of variation (regression)		**Source of variation (analysis of variance)**
Fit to regression line		Treatment variation
		Residual variation
<u>Residual variation</u>	—— becomes ⟶	<u>Total variation</u>
Total variation		

It is perfectly possible to have more than one independent variable in the analysis of covariance, and so to "correct" the experimental observations for two or more confounding variables. For example, in the potato experiment above, confounded by eelworm infestation, we may also have noticed that all the plots at the edge of the trial have suffered more slug damage than the other plots – again something we can measure and include as well as eelworm populations in the analysis of covariance.

EXECUTIVE SUMMARY 7
Linear regression

The **REGRESSION of _y_ on _x_** calculates the best fit straight line for predicting _y_ from measurements of _x_ : _y_ is usually represented on the vertical axis of a graph, and _x_ on the horizontal.

The slope (b) is how far you would rise (in _y_) for each unit of _x_, e.g. $15/20 = 0.75$ cm. Your height (y) at any horizontal distance (x) can be predicted as $y = bx$.

Visualize the regression slope as a flight of stairs and you will realise that halfway up the stair you are also halfway along horizontally. Where *mean y* and *mean x* intersect is one point on the regression line, and the coordinates in _y_ and _x_ units of the data points is measured as deviations from that unique location, The slope from there to any data point is therefore $(y - \bar{y})/(x - \bar{x})$.

The points on a graph do not usually show such a regular staircase, and each point will give a different answer for $(y - \bar{y})/(x - \bar{x})$. Our best estimate of the slope (b) will be to sum the information from all the points in some way.

The calculation of b using the "Sum of cross products"

$\sum(y - \bar{y})(x - \bar{x})$ is called the **sum of cross products**. Unlike the sum of squares of deviations (always positive), the sum of cross products can be positive or negative. How do we calculate it?

We can express a squared number as that number multiplied by itself, i.e. the sum of squares of _x_, i.e. $(\sum x^2 - (\sum x)^2/n)$, can equally be written as:

$$\sum(x \times x) - \frac{\left(\sum x \times \sum x\right)}{n}.$$

For the **sum of cross products** we simply replace one _x_ in each part by as _y_ as: $\sum(y \times x) - ((\sum y \times \sum x)/n)$. So for each point, we multiple _x_ and _y_, add these products together, and subtract the new correction factor based on the product of the two totals. _b_ is then:

$$\frac{sum\ of\ cross\ products}{sum\ of\ squares\ for\ x}.$$

The regression equation

We now have the slope (b), but usually _y_ is not zero when $x = 0$; i.e. the regression line crosses the _y_ axis at some intercept (a) above or below $y = 0$. Thus the full regression equation is **$y = a + bx$**, where _a_ is the height (y) at $x = 0$. We know the line passes through the coordinate $\bar{y}$ *with* $\bar{x}$ so we can substitute our means in the regression equation to solve the one unknown, a. The algebra for this is:

$$a = \bar{y} - b(\bar{x})$$

We know one point on the line is the coordinate $\bar{y}$ with $\bar{x}$. Find two other points by putting first a low and then a high value of x into $y = a + bx$. Then draw the line.

Analysis of variance of the regression

Each point on the graph can be measured on the y scale as $y - \bar{y}$. All these $y - \bar{y}$ values represent the "total variation." Each $y - \bar{y}$ is made up of two elements which together account for that variation:

1 Where the point OUGHT to be if it sat on the line at that particular position of $x - \bar{x}$. This is $\boldsymbol{b(x - \bar{x})}$.
2 The ERROR – how far the point is away on the y scale from where it ought to be.

We can easily work out the TOTAL VARIATION in y as sums of squares as $\sum y^2 - (\sum y)^2/n$ – our usual calculator method.

We can also work out how much of this is accounted for by (1) above – i.e. the variation of the points as if they all sat exactly on the regression line. This is summed for all the points from (1) above as:

$$\sum b[(x-\bar{x})^2], \text{ which is the same as } \boldsymbol{b^2} \times \textbf{sum of squares of } \boldsymbol{x}.$$

We now have TOTAL variation in y ($n - 1$ degrees of freedom) and how much of this is accounted for by the regression line (1 degree of freedom). The difference is the RESIDUAL ($n - 2$ degrees of freedom). We can then carry out the END PHASE to check that F (variance ratio) for the regression is significant – this tells us we have a significant relationship for y on x, i.e. b (slope) is significantly different from **zero** (no slope).

How well do the points fit the line?

We measure this by the PERCENTAGE of the total variation in y accounted for by the regression:

$$\frac{\textit{Sum of squares for regression}}{\textit{Sum of squares of } y} \times 100$$

This is called the "coefficient of determination." The statistic BEFORE multiplication by 100 is known as r^2. r, the correlation coefficient, is the square root of this value.

Spare-time activities

1 The data are measurements on 10 seeds of broad bean cv. *Roger's Emperor*

Weight (g)	Length (cm)
0.7	1.7
1.2	2.2
0.9	2.0
1.4	2.3
1.2	2.4
1.1	2.2
1.0	2.0
0.9	1.9
1.0	2.1
0.8	1.6

Draw a graph of the regression of weight (y) on length (x) of the bean seeds in order to use length as a simple quick guide to weight of the seed. Is the slope of the regression significant and what is the coefficient of determination (r^2)?

2 In an experiment on strawberry plants, the number of flower heads produced during the fruiting season was counted on individual plants, and then the eventual yield from those same plants was recorded. The data were as follows:

Number of flower heads	Yield (kg)
5	0.69
15	1.39
12	1.09
9	0.76
12	0.86
10	1.04
13	1.10
7	0.94
14	0.98
5	0.75
6	0.76
13	1.21
8	0.86
9	0.71
10	1.19

What is the linear regression equation to predict yield from the number of flower heads?
In order to be able to substitute a count of flower heads for measuring yield, about 80% of the variation should be accounted for by the regression. Do these data meet that criterion?

18

Chi-square tests

Chapter features

Introduction

I introduced the normal distribution early in this book (Chapter 5). It is the basis for judging what probability there is that an individual observation differing from the mean value by any given amount could be sampled by chance, given the mean and standard deviation of the distribution. I also explained that biologists usually accept a probability of less than once in 20 samples ($P < 0.05$) as being an event unlikely enough to suggest that the observation in question probably came from another normal distribution centred around a different mean value.

Statisticians have used the normal distribution to derive new statistics and provide significance tests based on distributions of one number divided by another. Thus the t-test uses the *difference between two means* divided by the *standard error of differences between two means*. The "F" test involves dividing one *variance* (called "mean square" in the analysis of variance) by *another* (usually the *residual mean square* in the analysis of variance). So tables of "t" and "F" exist which give the largest value of that statistic which chance sampling of one normal distribution would yield for different levels of probability and different "sizes" of experiments (as measured by degrees of freedom). Thus 1.96 is the largest chance value of the t statistic at $P = 0.05$ in very large experiments (∞ d.f., Table A2.1), whereas the equivalent largest chance value for F for two treatments (1 d.f. horizontally and ∞ d.f. vertically in Table A2.3) is a different number, 3.84.

In this chapter we meet yet another significance test employing a new statistic called "chi-square." "Chi" is the letter of the Greek alphabet denoted by the lower case symbol χ, so we represent the statistic "chi-square" by the statistical notation χ^2. This statistic was developed in the late 19th century by Karl Pearson, who used theoretical calculations from the normal distribution to work out the chance distribution of χ^2 at different degrees of freedom (Table A2.5). The ratio for which Pearson worked out the distribution may strike you as a little obscure – *it is the square of the difference between an observed number and the number your hypothesis suggested it should be* divided by that *"expected" number*. In notation (followed by English) we normally write the formula for χ^2 as:

$$\frac{(O - E)^2}{E} \quad \text{or} \quad \frac{(\text{Observed} - \text{Expected})^2}{\text{Expected}}$$

χ^2 is nearly always the summation of several values, so is perhaps better defined as:

$$\sum \frac{(O - E)^2}{E} \quad \text{or} \quad \sum \frac{(\text{Observed} - \text{Expected})^2}{\text{Expected}}$$

This statistic is actually extremely useful and figures quite often in the biological literature. I suspect this is partly because the calculation is so quick and simple; as a result, χ^2 is not infrequently misunderstood and misused.

When and where not to use χ^2

χ^2 is a test to use only with data in the form of **frequencies**. Thus you can test whether there are different proportions of species of herbaceous plants and shrubs in a new than in an ancient woodland, but you **cannot** use χ^2 to test whether the total number of species differs between the two habitats. You can test whether 235 men and 193 women is a good enough fit to a 50:50 sex ratio, but you cannot test whether 235 mice per sq km is a higher population than 193 mice per sq km at another site. There is of course an "expected value" of 214 mice for a hypothesis that there is no difference in mouse density between the sites, but in this example 235 and 193 are **not** frequencies.

So what is a **frequency**? Frequency data really can arise in only one way. It is the group of data in a sample which share a common characteristic (in the mice example above, 235 and 193 were **different** samples). You have to ask yourself: "Am I really partitioning a population into groups?" "Unpartitioned" populations are not frequencies. So, although you cannot test whether 235 and 193 mice are significantly different, you could test whether their sex ratio differs at the two sites. So never fall into the trap of using χ^2 for 1 d.f. to test whether two numbers are significantly different – in any case, just biologically it would be nonsense.

Percentages are also not frequencies. Frequencies have to be the actual experimental numbers. This is because χ^2 is "scale-dependent." The larger the numbers, the better the chances of the χ^2 test detecting a significant difference even where percentage differences remain unchanged.

For example, a 60% and 40% split can come from many frequencies e.g. 3 and 2, 6 and 4, 12 and 8, 24 and 16, 48 and 32, and so no. In this series the frequencies are doubling, but they are all 60:40% splits. We haven't yet talked about how we do the test, so take my word for it that the χ^2 values for the series above are respectively 0.8, 1.6, 3.2, 6.4, and 12.8. χ^2 is doubled each time we double the frequencies. Incidentally, for this series, any value of χ^2 more than 3.8 is significant at $P < 0.05$.

The problem of low frequencies

In the past, a lot has been made of whether a χ^2 test can be used if any of the frequencies involved are small. The traditional wisdom was that there should be no expected frequencies (i.e. the divisor(s) in the χ^2 equation) below 5, and that such a low frequency should be combined with the frequency in an adjacent class to eliminate the problem. Today, this is regarded as too restrictive, and frequencies should only be combined to eliminate expected values of less than one.

Yates' correction for continuity

By their very nature, observed frequencies tend to be integers and therefore "discontinuous," i.e. you may find frequencies of 4 and 5, but 4.2, 4.7, etc. are often just not possible. Yet the expected frequency in the formula for χ^2 may well be a continuous number. It is therefore often suggested that, if the frequencies are not continuous, then to play safe the observed integer is regarded as having been rounded up or down in the direction which would make the "Observed – Expected" part of the χ^2 formula a little too large. This difference (unless already less than 0.5) should therefore be reduced by 0.5 before squaring. The correction is named after the statistician who devised it and it is always called the "Yates" correction for continuity." I have to say this strikes me as odd – is it not really a correction for *discontinuity?*

In any case, it makes little difference unless the numbers are small. I tend only to use it for integer data less than 100 when only 2 frequencies are being compared (i.e. a χ^2 for 1 d.f.) or if some frequencies are less than 10.

The χ^2 test for "goodness of fit"

χ^2 tests enable us to judge whether observed frequencies deviate significantly from the frequencies expected on the basis of some hypothesis that we wish to test. The "goodness of fit test" applies to where each observation can be assigned to just one of a number of classes – for example, the frequencies of different eye color in a large sample of people. We could also apply it where we regard all those combining a particular eye color with a particular hair color as falling into a **single** class. The second test I shall mention (association χ^2) is where each observation has a place in each of two or more classifications. Then, we compare the observed frequencies with the frequencies expected on the hypothesis that there is no association between the classifications, i.e. the observed frequencies are purely random. For example, we could secondarily divide the frequencies of each eye color into males and females and ask "are any eye colors more common in one sex than the other?"

But back to *goodness of fit*. We'll begin with a classification of our observations into two classes only. In rearing some biological control agents related to wasps, the danger is known that an apparently flourishing culture may suddenly crash because overcrowding has led to few females being produced by a species which normally produces equal proportions of males and females. So samples are regularly taken from such cultures to check that no statistically significant switch to a male biased sex ratio has occurred.

Say we have collected 98 adults and found them to comprise 54 males and 44 females. Should we be getting worried? Well, the 50:50 expectation is clearly $98/2 = 49$ individuals in each sex. (Observed – Expected)^2 for males is $(54 - 49)^2 = 5^2$. However, with just 2 frequencies, this is a case for correction for continuity. So we subtract 0.5 from the difference and use $4.5^2 = 20.25$. For females it is equally $(44 - 49)^2 = -5^2$, which is also 25 since the square of a minus value is a plus. Combining the χ^2 for both sexes gives:

$$\sum \frac{\text{Observed} - \text{Expected} - 0.5)^2}{\text{Expected}} = \frac{20.25}{49} + \frac{20.25}{49} = 0.41 + 0.41 = 0.82$$

How many degrees of freedom are appropriate in this case? Not difficult – two classes (male and female) gives $2 - 1 = 1$ d.f.

So we need to look up in the table of χ^2 (Table A2.5) what is the value χ^2 needs to attain at $P = 0.05$ and 1 d.f. It is 3.84. We do **not** have confirmation that the sex ratio has become significantly biased towards males; our sample has a greater than 1 in 20 chance of having been taken from a population where males and females are in equal numbers.

Since the "expected" frequencies are a re-apportioning (to reflect the hypothesis being tested) of the observed total, the **totals** of the observed and the expected frequencies are always identical.

The expected hypothesis does **not** have to be one of equality. Good examples of this come from the field of genetics, since there are known ratios of characteristics defining certain forms of inheritance. Some of these ratios stem from the famous work of the monk Gregor Mendel late in the 19th century, and I'll use his characteristics of garden peas with χ^2 to develop this part of the chapter further.

After crossing parents with normal green round peas, the pods of the offspring show peas which differ in two characteristics – the color of the seed coat (green or yellow) and the structure of the seed coat (round or wrinkled). A sample of 896 peas gives the following frequencies:

green & round (GR)	green & wrinkled (GW)	yellow & round (YG)	yellow & wrinkled (YW)
534	160	165	37

The parents carried all these characters, but that yellow and wrinkled only appeared in their offspring suggests that these characters are recessive as compared with dominant green and round.

Mendel's work concluded that green and yellow should occur in the ratio 3:1. So of the 896 peas, $3/4$ (i.e. 672) are expected to be green and $1/4$ (i.e. 224) to be yellow. The observed numbers are $534 + 160 = 694$ green and $165 + 37 = 202$ yellow. Do these frequencies fit Mendel's 3:1 ratio?

So the χ^2 test goes:

	Green	Yellow
Observed	694	202
Expected	672	224
Observed – Expected	22	−22
(Observed – Expected)^2	484	484
(Observed – Expected)^2/ Expected	$484/672 = 0.72$	$484/224 = 2.16$

$$\chi^2_{(P=0.05,1\ \text{d.f.})} = \qquad 0.72 \qquad + \qquad 2.16 \qquad = \mathbf{2.88}$$

d.f. = n − 1 for 2 classes = 1

In the table, χ^2 for $P = 0.05$ and 1 d.f. is 3.84, so our pea color ratio fits the expected 3:1 ratio. The other character of seed coat structure should similarly fit a 3:1 ratio. Does it?

	Round	Wrinkled
Observed	$534 + 165 = 699$	$160 + 37 = 197$
Expected (as before, 3:1)	*672*	*224*
Observed – Expected	27	−27
(Observed – Expected)^2	729	729
(Observed – Expected)^2/Expected	$729/672 = 1.08$	$729/224 = 3.25$
$\chi^2_{(P=0.05,1 \text{ d.f.})} =$	1.08 +	3.05 $= 4.33$

This is now larger than the 3.84 in the table: the seed coat structure does **not** fit the expected 3:1 ratio, but see below for further statistical detective work!

The case of more than two classes

We can get a better idea of what is going on by returning to the frequencies in all four pea categories, and comparing the observed frequencies with the ratio of 9:3:3:1 that Mendel suggested would apply with the combination of two characters (one dominant and the other recessive). After the earlier examples above, I hope it is OK for me to leave out some of the detail of the calculations; they can be found in Box 18.1 (if needed).

The sum of the expected ratios $9 + 3 + 3 + 1 = 16$, so the expected frequencies will be calculated in the relevant "sixteenths" of the total of 896 seeds:

	GR	GW	YR	YW
Observed	534	160	165	37
Expected	$896 \times 9/16$	$896 \times 3/16$	$896 \times 3/16$	$896 \times 1/16$
	$= 504$	$= 168$	$= 168$	$= 56$
Difference (O – E)	30	−8	−3	−19
$\chi^2 (O-E)^2/E$	1.78 +	0.38 +	0.05 +	6.45 $= \mathbf{8.66}$

Four categories give 3 d.f. and in the table $\chi^2_{(P=0.05,3 \text{ d.f.})}$ is 7.81. The observed frequencies deviate significantly from a 9:3:3:1 ratio.

You will note that, with more than two classes, the χ^2 test has the similarity with the F-test (in the analysis of variance) that the existence of a significant difference is identified – but not which frequency or frequencies differ(s) significantly from their expected one. They might all do so, or only one.

BOX 18.1

In this box regarding χ^2 calculations, observed frequencies are shown in **BOLD** and expected frequencies in **_BOLD ITALICS_**.

Example of goodness-of-fit to a 9:3:3:1 ratio (page 282):

$$\chi^2 = \frac{(\mathbf{534} - \mathbf{504})^2}{\mathit{504}} + \frac{(\mathbf{160} - \mathit{168})^2}{\mathit{168}} + \frac{(\mathbf{165} - \mathit{168})^2}{\mathit{168}} + \frac{(\mathbf{37} - \mathit{56})^2}{\mathit{56}}$$

Example of goodness-of-fit to a 9:3:3 ratio (page 284):

$$\chi^2 = \frac{(\mathbf{534} - \mathit{515.4})^2}{\mathit{515.4}} + \frac{(\mathbf{160} - \mathit{171.8})^2}{\mathit{171.8}} + \frac{(\mathbf{165} - \mathit{171.8})^2}{\mathit{171.8}}$$

Example of limpets and black and white backgrounds (total frequencies):

$$\chi^2 = \frac{(\mathbf{71} - \mathit{59.5} - 0.5)^2}{\mathit{59.5}} + \frac{(\mathbf{48} - \mathit{59.5} - 0.5)^2}{\mathit{59.5}}$$

Example of limpets and black and white backgrounds (by individual board):

Frequencies	Expected (equality)	χ^2 (with correction for continuity)
18 and 3	10.5	$\dfrac{(\mathbf{18} - \mathit{10.5} - 0.5)^2}{\mathit{10.5}} + \dfrac{(\mathbf{3} - \mathit{10.5} - 0.5)^2}{\mathit{10.5}}$
20 and 8	14.0	$\dfrac{(\mathbf{20} - \mathit{14.0} - 0.5)^2}{\mathit{14.0}} + \dfrac{(\mathbf{8} - \mathit{14.0} - 0.5)^2}{\mathit{13.0}}$
7 and 16	11.5	$\dfrac{(\mathbf{7} - \mathit{11.5} - 0.5)^2}{\mathit{11.5}} + \dfrac{(\mathbf{16} - \mathit{11.5} - 0.5)^2}{\mathit{11.5}}$
10 and 11	10.5	$\dfrac{(\mathbf{10} - \mathit{10.5} - 0.5)^2}{\mathit{10.5}} + \dfrac{(\mathbf{11} - \mathit{10.5} - 0.5)^2}{\mathit{10.5}}$
16 and 10	13.0	$\dfrac{(\mathbf{16} - \mathit{13.0} - 0.5)^2}{\mathit{13.0}} + \dfrac{(\mathbf{10} - \mathit{13.0} - 0.5)^2}{\mathit{13.0}}$
71 and 48	59.5	$\dfrac{(\mathbf{71} - \mathit{59.5} - 0.5)^2}{\mathit{59.5}} + \dfrac{(\mathbf{48} - \mathit{59.5} - 0.5)^2}{\mathit{59.5}}$

Really the only technique for sorting this out is to run the χ^2 calculation again with the class which appears the most deviant from the expected ratio excluded. If the remaining classes now fit the remaining expected ratio, one can be satisfied that it is this class which has caused the significant total χ^2 value. If the ratio is still not fitted, there is clearly a second deviating class, and the calculation is re-run with the remaining most deviant class excluded – and so on.

Clearly the YW class produces the largest contribution (6.45) to the total χ^2 of 8.66. We therefore exclude this class and test the fit of the remaining three classes to what is left of the expected ratio, i.e. 9:3:3. These ratios now adding up to 15 with 859 seeds remaining, making 9/15 the proportion of the total number of seeds that are expected to be GR:

	GR	GW	YR
Observed	534	160	165
Expected	859 × 9/15	859 × 3/15	859 × 3/15
	= 515.4	= 171.8	= 171.8
Difference (O − E)	18.5	−11.8	−6.8
χ^2(O − E)^2/E	0.66 +	0.81 +	0.27 = **1.74**

The total χ^2 (now for $3 - 1 = 2$ d.f.) is well below the table value of 5.99 at $P = 0.05$. Clearly GR, GW, and YR **do** fit the Mendelian ratio of 9:3:3:1 well, but the YW class has significantly fewer seeds than expected. This is the double recessive combination, which may well sometimes fail to produce a seed.

Quite a nice story and also a neat example of using χ^2.

χ^2 with heterogeneity

In the first test in this chapter, a test which concerned sex ratios, we tested the fit of observed frequencies to a 50:50 expected male:female ratio. Statisticians seem to have no problem with two unreplicated frequencies, but χ^2 tests for 1 d.f. have always bothered me as a biologist. OK, I guess I accept testing the fit to a 50:50 ratio substantiated by genetics theory, but is this really different from testing whether two observed frequencies differ from an assumption of equality?

Suppose we put a black and white checkerboard in of each of five tanks of seawater and toss 30 limpet molluscs into each tank. We record whether they have settled on black or white squares, ignoring those who have settled across a boundary. Out of the 150 limpets, we find 71 have settled on black squares and 48 on white ones. The expectation assuming no preference is of course $(71 + 48)/2 = 59.5$. χ^2 for 71 and 48 with an expected frequency of 59.5 (and correction for continuity as there are only two frequencies and limpet numbers are bound to be integers) is 4.07 (see Box 18.1 for the full calculation if you need it), above the 3.84 for $P = 0.05$ and 1 d.f. in the χ^2 table. This might convince us that limpets prefer to settle on dark surfaces. But hang on a minute! One of our five checkerboards (see below) has only seven limpets on black squares compared with more than twice as many (16) on white. Also, another checkerboard has almost the same frequency on both colors (10 and 11). So, in spite of the apparent large difference between 71 and 48, the evidence that limpets prefer to settle on dark surfaces is not that convincing after all, is it?

The results for all five boards are:

Black	White
18	3
20	8
7	16
10	11
16	10

A heterogeneity χ^2 analysis includes a statistical measure of how consistently the overall difference is reflected by the different replicates, guiding us as to how far a significant χ^2 between the two overall totals can be trusted.

To do such an analysis, we first calculate a separate χ^2 for 1 d.f. for each of the five boards, and total them to arrive at a χ^2 for 5 d.f. Correction for continuity is again needed (see earlier) and so we subtract 0.5 from *Observed – Expected* differences (again Box 18.1 has the calculations involved):

Black	White	$\chi^2_{(1 \text{ d.f.})}$
18	3	9.33
20	8	2.70
7	16	2.78
10	11	0.00
16	10	0.96
	$[\chi^2_{(5 \text{ d.f.})} =$	**15.77]**
Overall 71	48	**4.07** (with correction for continuity)

The heterogeneity χ^2 analysis table is not dissimilar from a simple analysis of variance. The χ^2 values for the five replicates total 15.77 for 5 d.f., of which 4.07 is accounted for by the overall result (71 *v.* 48) for 1 d.f., leaving a "residual" (by subtraction) with $5 - 1 = 4$ d.f. as an estimate of the lack of consistency (= heterogeneity) of the overall result across the five replicates. The table is therefore as follows, with the appropriate P values from the χ^2 table:

Source of χ^2	d.f.	χ^2	P
Overall (71 v. 48)	1	4.07	<0.05
Heterogeneity (by subtraction)	4	11.70	<0.05
Total	5	15.77	

You can see that although the χ^2 for the overall distribution to 1 d.f. of the limpets between black and white backgrounds is significantly high, so is the χ^2 for heterogeneity. The conclusion, based on the χ^2 for 1 d.f. that the limpets have settled more on black than white backgrounds, has to be regarded as unsafe in spite of the apparent large difference. The χ^2 for heterogeneity has picked out that there is significant inconsistency between the replicates in how the limpets are distributed. The advantage of replicating experiments so as to make them amenable to a heterogeneity χ^2 analysis is clear.

Heterogeneity χ^2 analysis with "covariance"

Well, it's not really covariance, but I didn't know what else to call it! There are sometimes reasons why expected values for different replicates may vary because of some additional effect which affects the replicates differentially. I guess this would make more sense with an example?

Say we want to assess the preference of aphids for two different varieties of cabbage (we'll call them A and B). An approach to doing this might be to confine a number of aphids in a cage enclosing equal areas of leaf of each variety (potted plants). To know how many aphids have really settled down to feed on either leaf portion, we decide to treat the pot of one plant of each pair with a systemic insecticide. This will poison the plant sap and kill any aphid that imbibes that sap, even if only for a short time. It makes sense to have an even number of replicates, in half of which variety A is treated with insecticide and B is treated in the other replicates. The results of 10 replicates might be as follows (dead aphids are assumed to have preferred the insecticide-treated variety):

Replicate	Dead	Alive
Variety A treated		
1	16	4
2	15	5
3	18	2
4	20	0
5	14	6
Variety B treated		
6	12	8
7	10	10
8	13	7
9	9	11
10	13	7
Totals dead and alive	**140**	**60**

If you look at which variety was treated in each replicate, the figures in italics (totalling 126) can be presumed to be aphids which have settled to feed on variety A (since they have encountered the poison applied to A and avoided that applied to B) and the regular typeface aphids (total 74) can be presumed to have fed on variety B (by the reverse argument). The $\chi^2_{(1 \text{ d.f.})}$ for 126 v. 74 is significantly high (see later), indicating the aphids have shown us a preference for Variety A, but there are another two totals to consider in this experiment! These are the totals of 140 dead and only 60 alive aphids in the experiment. If the insecticide had no effect on how the aphids are distributed, then we would expect both totals to be 100, since in both varieties there were as many untreated as treated plants. Clearly, regardless of which variety received insecticide in each replicate, more aphids settled on whichever variety in the replicate was treated with insecticide and were poisoned. Presumably, many aphids tried feeding on both varieties at some time.

In each of the ten replicates, therefore, the expected values for the heterogeneity χ^2 analysis (assuming **no** varietal preference) are 14 dead per replicate for the insecticide-treated variety and six alive per replicate for the untreated variety.

The heterogeneity χ^2 analysis therefore becomes (again see Box 18.2 for more detailed formulae):

Replicate	Variety A Observed	Variety A Expected	Variety B Observed	Variety B Expected	$\chi^2_{(1 \text{ d.f.})}$
Variety A treated					
1	16	14	4	6	0.54
2	15	14	5	6	0.06
3	18	14	2	6	2.92
4	20	14	0	6	7.20
5	14	14	6	6	0.00
Variety B treated					
6	8	6	12	14	0.54
7	10	6	10	14	2.92
8	7	6	13	14	0.06
9	11	6	9	14	3.49
10	7	6	13	14	0.06
					$[\chi^2_{(10 \text{ d.f.})} = \textbf{17.79}]$
Overall	**126**	**100**	**74**	**100**	**13.00**

(with correction for continuity)

BOX 18.2

Example of aphid plant variety preference test using insecticide:

Frequencies (expected frequency in brackets)	χ^2 (with correction for continuity)
16 (14) and 4 (6)	$\dfrac{(16-14-0.5)^2}{14} + \dfrac{(4-6-0.5)^2}{6}$
15 (14) and 5 (6)	$\dfrac{(15-14-0.5)^2}{14} + \dfrac{(5-6-0.5)^2}{6}$
18 (14) and 2 (6)	$\dfrac{(18-14-0.5)^2}{14} + \dfrac{(2-6-0.5)^2}{6}$
20 (14) and 0 (6)	$\dfrac{(20-14-0.5)^2}{14} + \dfrac{(0-6-0.5)^2}{6}$
14 (14) and 6 (6)	$\dfrac{(14-14-0.0)^2}{14} + \dfrac{(6-6-0.0)^2}{6}$
8 (6) and 12 (14)	$\dfrac{(8-6-0.5)^2}{6} + \dfrac{(12-14-0.5)^2}{14}$
10 (6) and 10 (14)	$\dfrac{(10-6-0.5)^2}{6} + \dfrac{(10-14-0.5)^2}{14}$
7 (6) and 13 (14)	$\dfrac{(7-6-0.5)^2}{6} + \dfrac{(13-14-0.5)^2}{14}$
11 (6) and 9 (14)	$\dfrac{(11-6-0.5)^2}{6} + \dfrac{(9-14-0.5)^2}{14}$
7 (6) and 13 (14)	$\dfrac{(7-6-0.5)^2}{6} + \dfrac{(13-14-0.5)^2}{14}$
126 (100) and 74 (100)	$\dfrac{(126-100-0.5)^2}{100} + \dfrac{(74-100-0.5)^2}{100}$

The heterogeneity χ^2 table is therefore:

Source of χ^2	d.f.	χ^2	P
Overall (126 v. 74)	1	13.00	<0.001
Heterogeneity (by subtraction)	8	4.79	n.s.
Total	9	17.79	

The heterogeneity χ^2 is extremely small (at $P = 0.05$ and 8 d.f. it would have to be at least 15.51), and so we can readily accept the highly significant overall χ^2 as evidence that the aphids preferentially feed on Variety A.

Association (or contingency) χ^2

This is a rather different use of χ^2, though the basic *(Observed – Expected)^2/Expected* calculation, the requirement for frequencies, the problem of small expected values, and the correction for continuity all still apply. Association χ^2 tests involve multiple categorization of frequencies. The concept of association is found in simple questions like "Is red-green color-blindness associated with one gender rather than the other?"

Each individual in the test is not only either color-blind (CB) or not (Normal), but is also either a man or a woman. In goodness of fit tests we had a linear series of frequencies:

Category	CB men	Normal men	CB Women	Normal women
Frequency	44	18	10	39

In this example, what are the expected values? We have no ratio of frequencies we can test for goodness of fit, and 1:1:1:1 would be stupid, if only because more men than women were tested for color vision.

This is where we need an association test with its multidimensional table (called the contingency table).

2 × 2 contingency table

For the above data on color-blindness, the four data would form a table with two rows and two columns (i.e. a 2 × 2 table):

	Color-blind	Normal	**Totals**
Men	44	18	62
Women	10	39	49
Totals	54	57	**111**

We have 111 folk, of whom 44 are both men and color-blind. Given no association between men and color-blindness, we would expect the proportion of color-blind men and women to be identical , i.e. about half (54 out of 111). With 62 men, there should therefore be $62 \times (54/111) = 30.2$ who are color-blind, but at 44 the frequency is rather higher, suggesting some association between color-blindness and gender. Similarly, the expected color-blind women is $49 \times (54/111) = 23.8$, and the expected frequencies of normal men and women respectively $62 \times (57/111) = 31.8$ and $49 \times (57/111) = 25.2$. Of course, just as with the goodness-of-fit test, these four expected frequencies of 30.2, 23.8, 31.8, and 25.2 again add up to the same grand total of 111 for the number of people tested.

You'll see that all the expected frequencies are the respective row and column totals multiplied together and then divided by the grand total. Simple to remember, I think. Is this a familiar calculation? Probably not, but we have been there before in the analysis of variance (Box 11.2) when working out the expected values of data if there were no interaction.

So we can now add the "expected" values (in brackets and italics) to our contingency table:

	Color-blind	Normal	**Totals**
Men	44 *(30.2)*	18 *(31.8)*	62
Women	10 *(23.8)*	39 *(25.2)*	49
Totals	54	57	**111**

This gives us four χ^2 values to calculate and add to obtain an overall χ^2, but how many d.f. does it have? Well, you might think four numbers gives us 3 d.f., but think again of the analogy with interactions (page 150). There is just 1 degree of freedom. Put in any one figure (say 10 for color-blind women), and the other three numbers follow inevitably from the column and row totals:

	Color-blind	Normal	**Totals**
Men	54–10	62–(54–10)	62
Women	10	49–10	49
Totals	54	57	**111**

So remember, degrees of freedom for contingency tables are (columns − 1) × (rows − 1) = 1 × 1 = 1 in the above example.

The calculation of the four χ^2 values proceeds as before; the detailed calculations are given in Box 18.3. We can then insert these values in the table (in bold and in square brackets):

	Color-blind	Normal	**Totals**
Men	44 *(30.2)* **[6.31]**	18 *(31.8)* **[5.99]**	62
Women	10 *(23.8)* **[8.00]**	39 *(25.2)* **[7.56]**	49
Totals	54	57	**111**

BOX 18.3

In this box regarding χ^2 calculations, observed frequencies are shown in **BOLD** and expected frequencies in ***BOLD ITALICS***.

Example of color-blindness in men and women (page 289):

For color-blind men, $\chi^2 = \dfrac{(44 - 30.2)^2}{30.2} = 6.31$

For color-blind women, $\chi^2 = \dfrac{(10 - 23.8)^2}{23.8} = 8.00$

For normal men, $\chi^2 = \dfrac{(18 - 31.8)^2}{31.8} = 5.99$

For normal women, $\chi^2 = \dfrac{(39 - 25.2)^2}{25.2} = 7.56$

Example of cyanogenic bird's foot trefoil at coast and inland (page 292):

For +CG at coast, $\chi^2 = \dfrac{(17 - 25.5)^2}{25.5} = 2.83$

For +CG inland, $\chi^2 = \dfrac{(48 - 39.5)^2}{39.5} = 1.83$

For −CG at coast, $\chi^2 = \dfrac{(14 - 10.2)^2}{10.2} = 1.42$

For −CG inland, $\chi^2 = \dfrac{(12 - 15.8)^2}{15.8} = 0.91$

For Both at coast, $\chi^2 = \dfrac{(45 - 40.4)^2}{40.4} = 0.52$

For Both inland, $\chi^2 = \dfrac{(58 - 62.6)^2}{62.6} = 0.34$

Adding these four χ^2 values together gives the highly significant ($P < 0.001$) total χ^2 for 1 d.f. of 27.86. Inspection of observed and expected values leads to the interpretation that red-green color-blindness is significantly commoner in men than in women.

Fisher's exact test for a 2 × 2 table

I pointed out earlier that low expected frequencies are a problem with χ^2 tests. In a 2 × 2 table, a single small expected frequency of less than 5 represents a large proportion of the data, and the famous pioneer statistician R.A. Fisher devised his "Exact test" for such situations. This involves the factorial values of the numbers in the table. Remember factorials (Box 14.1)? – not the experimental design which bears that name but the function of a number (e.g. 7! denotes the factorial value of the number 7). To remind you, in mathematics the "factorial" is the product of a descending series of numbers beginning with the number to be factorialized – e.g. factorial 7 or 7! is $7 \times 6 \times 5 \times 4 \times 3 \times 2 \times 1 = 5040$.

Fisher's "Exact test" is beyond the scope of this book. You will find details in larger statistical texts, but the following will give you a rough idea of what's involved. First the factorials of all the column and row totals are multiplied together and then the result is divided by another product of factorials using the grand total and the numbers within the table. This is repeated with the smallest number reduced by one (but no totals changed as a result), with the final repeat when it reaches zero. The "Exact test" does not produce a χ^2 of which the probability has to be found in tables; instead, the P value is found directly by adding together the result of the several repeat calculations.

Larger contingency tables

There is no limit to the number of columns and rows in a contingency table; each time the degrees of freedom for the resulting χ^2 will be (columns − 1) × (rows − 1).

As an example of a contingency table larger than 2×2, I'll use a 3×2 table. My example relates to a common leguminous plant, the "bird's foot trefoil." One form of two forms of this plant produces cyanogenic glucosides poisonous to many herbivorous animals; the other form does not. The data below show results from a presence or absence study of the two forms in a large number of sampled $0.5\,\text{m} \times 0.5\,\text{m}$ areas in two locations, at the coast and at least 3 km inland. Squares were categorized as containing only the cyanogenic form (+CG), only the acyanogenic form (−CG), or some of each form (Both):

	+CG	−CG	Both	**Totals**
Coast	17	14	45	76
Inland	48	12	58	118
Totals	65	26	103	**194**

We now have six expected values to calculate by the formula I explained earlier for contingency tables: expected value = row total × column total/grand total. So for +CG at the coast, the expected value is

$76 \times 65/194 = 25.5$. The other expected values are calculated similarly, and added to the table in italics and round brackets:

	+CG	−CG	Both	**Totals**
Coast	17 *(25.5)*	14 *(10.2)*	45 *(40.4)*	76
Inland	48 *(39.5)*	12 *(15.8)*	58 *(62.6)*	118
Totals	65	26	103	**194**

The calculation of the six χ^2 values proceeds as before, the detailed calculations are given in Box 18.3. We can then again insert these values in the table (in bold and in square brackets):

	+CG	−CG	Both	**Totals**
Coast	17 *(25.5)* **[2.83]**	14 *(10.2)* **[1.42]**	45 *(40.4)* **[0.52]**	76
Inland	48 *(39.5)* **[1.83]**	12 *(15.8)* **[0.91]**	58 *(62.6)* **[0.34]**	118
Totals	65	26	103	**194**

The total of these six χ^2 values is 7.85. Degrees of freedom = (columns − 1) × (rows − 1) = 2 × 1 = 2. Table A2.5 in Appendix 2 tells us that a χ^2 value as high as 7.85 at 2 d.f. is statistically significant ($P < 0.05$).

How should we interpret the results of this survey biologically? Well, the largest contributions to the χ^2 of 7.85 come from the distribution of +CG plants (which are more frequent than expected inland and less than expected at the coast) and the greater than expected frequency of −CG at the coast. But this doesn't explain the biology of what is happening. Surveys like this are not experiments, but can be used to show that frequencies are at least consistent with a hypothesis, even though they cannot prove it. In this case the hypothesis being statistically supported is that the distribution of cyanogenesis in the plants reflects the distribution of snails, which are deterred by the cyanogenic compounds from feeding on the +CG form of bird's foot trefoil. Thus, inland, the unprotected −CG plants are at a disadvantage compared with +CG plants. However, snails cannot tolerate the salty environment at the coast, so the −CG plants are no longer disadvantaged and their frequency in comparison with +CG plants increases.

Interpretation of contingency tables

We have just seen how a contingency table can be used to support a hypothesis. But it is dangerous to use the data to try and create a hypothesis!

The obvious deduction from the data is the direct one that –CG plants tolerate saline soils better than +CG ones – you have to know the snail connection to interpret the contingency table correctly.

So, rather like regression (page 239), contingency tables can identify phenomena – but they cannot explain them. For that you need an experiment with treatments specifically designed to identify explanations. By contrast, the contingency table may appear to answer a question, but it just as well answers other questions, some of which may paint a totally different picture.

Take a contingency table which divides imprisoned and tagged convicts into those that do and do not re-offend, and which gives frequencies that might convince a social worker that tagging offenders rather than putting them in prison reduces the risks of a repeat offence. A more cynical prison officer might, however, use the identical contingency table to argue that tagging is applied mainly to first offenders committing less serious crimes, while the hardened criminals who regularly re-offend go straight to prison without the tagging option being considered.

Spare-time activities

1 In sampling aphids, many workers turn over plant leaves and count the number of aphids they see in the different growth stages (instars). It takes a lot of time, and a lot of skill to identify the instars! But is the method accurate?

Here's one such field count (in the column headed "Inspection"). The second column was obtained by taking some other leaves back to the laboratory and washing the aphids off into a sieve under a jet of water – this effectively cleans everything off the leaves and therefore gives a very accurate count:

Instar	Inspection	Washing
1	105	161
2	96	114
3	83	85
4	62	63
Adult	43	47
Total	**389**	**470**

Use χ^2 to determine how well the inspection counts reflect the true instar distribution? Is there any instar which appears particularly poorly estimated by inspection?

Note: The expected values of the observed "inspection" counts will be their total (389) divided up to reflect the proportions found in the "washing" counts.

2 On different evenings, groups of 10 snails were released onto a small tray fixed onto a brick wall. The next morning, the snails which had moved at least 10 cm upwards or downwards from the tray were counted:

Trial	Upwards	Downwards
1	7	3
2	5	4
3	4	4
4	6	4
5	5	5
6	7	2

Use a heterogeneity χ^2 test to analyze whether there is a difference in the frequency of upwards versus downwards movement.

3 The larvae of a fly tunnel within holly leaves, and suffer mortality from two main sources – birds peck open the mine and eat the insects, and the larvae of small wasps (parasitoids) feed internally on the fly larvae and then pupate within the leaf mine.

   The fate of the larvae in 200 leaves picked from the top, middle, and bottom of the tree was as follows:

	Top	Middle	Bottom
Healthy larva	9	41	18
Healthy pupa	11	23	6
Parasitoid pupa	8	19	24
Eaten by bird	15	20	6

Can you confirm that the frequency of either of the mortality factors is greater at one tree stratum than the others?

19

Nonparametric methods (what are they?)

Chapter features

Disclaimer

I'd better come clean right at the start of this chapter, and explain that I'm not going to try and teach you any of the nonparametric methods of data analysis. Each one has a unique formula for calculating its own statistic (examples are given in Box 19.1), and each then requires its own tables for testing the significance of that statistic. This book is limited to the most widely used traditional parametric methods. I'm not competent to explain the theory behind the different nonparametric methods, and so I would not be able to add anything to the recipes and tables that are already widely available, including on the internet. I'm therefore limiting my sights to where I think I can contribute to your understanding. I will introduce you to the idea of nonparametric methods, to explain how they differ from parametric ones, and where they may be particularly useful. Finally, I will guide you towards those nonparametric methods able to substitute for some of the parametric methods featured in this book.

Introduction

This book has so far dealt only with parametric methods, by which is meant methods that fit our data to theoretical distributions. Such distributions can be summarized by statistical *parameters* such as average, variance, standard deviation, regression coefficients, etc. "Parameter" is Greek for "beyond measurement," so I always allow myself a smile whenever I read

---

BOX 19.1

The statistic H of the Kruskal–Wallis analysis of ranks from the example on page 302

The two figures needed are n, which is the number of observations $= 12$ and $\sum R$, the sum of squared "treatment" totals first divided by number of data in the treatment

$$= \frac{32.00^2}{4} + \frac{36.00^2}{4} + \frac{10.00^2}{4} = 256.00 + 324.00 + 25.00 = 605$$

The equation for H is:

$$\left(\sum R \times \frac{12^*}{n(n+1)} \right) - 3^*(n+1) = \left(605 \times \frac{12^*}{12(3)} \right) - 3^*(13)$$

$$= (605 \times 0.077) - 39 = 7.585$$

*12 and 3 are constants and have nothing to do with, for example, e.g. that $n = 12$ in this example.

Note also that the divisors in the calculation of $\sum R$ need not be identical, so that treatments with different numbers of replicates can be accommodated.

The statistic r, Kendall's rank correlation coefficient, from the example on page 303.

The two figures needed from page 303 are $\sum(P - Q) = +38$ and the number of pairs, $n = 10$.

The equation for Kendall's r is:

$$\frac{2 \times \sum(P - Q)}{n(n-1)} = \frac{+76}{10(11)} = \frac{+76}{110} = +0.691$$

"The following parameters were measured." No, in parametric statistics we recognize that, although the theoretical distributions on which we base our statistical analyzes are defined by parameters, we have to work with samples from real populations. Such samples give us, with varying accuracy, only **estimates** of the parameters that are truly "beyond measurement."

Parametric statistics analyze the distances between numbers and the mean or each other, and fit these distances to theoretical distributions (particularly the normal distribution). Estimates of the parameters which can be derived from the theoretical distribution (such as the s.e.d.m.) are used to calculate the probability of other distances (e.g. between means of two treatments) arising by chance sampling of the single theoretical population. Parametric methods are therefore essentially quantitative. The absolute value of a datum matters.

By contrast, nonparametric methods are more qualitative. The absolute value of a datum is hardly important – what matters is sometimes merely whether it is larger or smaller than other data. So means of 0.8, 17, and 244 can be "equal" nonparametrically in all being smaller than a mean of 245!

Thus nonparametric statistics are all about the probability of an event occurring without it relating to any quantitative distribution. To judge if a lettuce of a certain weight is improbably larger involves measuring other lettuces also, but to say that the probability of throwing 3 sixes in sequence with a dice is $1/6 \times 1/6 \times 1/6 =$ about 0.005 (i.e. once in 200 goes) is a nonparametric calculation – no actual quantitative measurements are needed to calculate that probability.

Advantages and disadvantages of the two approaches

Where nonparametric methods score

The big disadvantage of parametric methods is their basis in quantitative theoretical distributions – if the data don't fit the theoretical distribution, then a parametric test is likely to lead to the wrong conclusions. So important assumptions in parametric statistics are that data are normally distributed and that the variability within the populations being compared is essentially similar. Thus, in the analysis of variance, standard deviations in the different treatments should be similar in spite of different means (page 103). This is often hard to tell if each treatment has only a few replicates! Biologists tend to assume the data **are** normally distributed, though you would probably need at least 20 replicates of a treatment to be sure. But we often only have three or four. Should we therefore give up parametric tests? We need to be sensible about this. We would expect many biological characteristics to show individual variation of a "normal" type, with most replicate samples close to the mean, and more extreme values increasingly less frequent in both directions away from that mean. It is often said that one cannot test the assumption of normality with few samples, and we should indeed inspect our data to check as far as we can that no obvious departure from normality is likely. We may settle for normality on theoretical grounds or from previous experience of the type of datum, personal or from the literature. A useful check possible with even just a few replicates of several treatments, is to look for the tell-tale sign of trouble that standard deviations or errors increase with the size of those means (page 37). Transformation of the data (page 38) may then be a solution allowing parametric methods nonetheless to be used.

Things are very different in medical statistics. Clearly the incidence of an unusual disease will not be normally distributed in a population but target some age or genetic groups more than others. Moreover, if medical treatments are any good, some treatments can have very little standard deviations compared with others (e.g. the top treatment may give close to a 100% cure with very little variation). It is therefore perhaps not surprising that nonparametric tests are used far more widely in medicine than in biology.

Unfortunately, cleverly designed though computer software often is, the human still has to decide whether a parametric or a nonparametric test is appropriate. Some data can be analyzed both ways – some only by nonparametric methods. One example of the latter is where we know the different populations of numbers we are comparing have very different distributions, as in the contrast between many zeros and only few insects where we have applied an insecticide and a wide distribution of insect numbers on untreated plants. Another example is where the data are *ordinal*. This is where the data are ranks and not the original numbers which have been ranked. Thus we might have ranked a school class from first to last on both their French and mathematics marks and wish to test how far the same students are good at both subjects. That nonparametric tests frequently use ranks or simply direction of differences, can make them quick and simple to use. The collection of data can also be far less time-consuming than that for parametric tests.

Where parametric methods score

Data suitable for parametric tests can often also be analyzed by a nonparametric method. The parametric method then has several advantages. The significance of interactions in factorial experiments and the relative importance of the factors being factorialized can be identified, whereas in analogous nonparametric tests, "treatments" have to be just "treatments" even when they are a combination of levels of more than one factor. In other words, with nonparametric tests we have no Phase 3 (page 158) for the analysis of factorial experiments. Parametric methods also have the advantages of greater power and greater ability to identify between which treatment means there are significant differences. As regards greater power, the parametric test will be less likely to reject a real difference than a nonparametric test. Thus, often data suitable for the parametric "*t*-test" could equally be analyzed by the nonparametric "sign test," but you would need to do more sampling with the sign test to achieve the same level of discrimination between the means as with the *t*-test. When the sample size is small, there is little difference between the two tests – only 5–6% more samples need to be taken in comparison to a *t*-test. But the difference in the relative power of the two tests does increase with number of samples, and rises to close on 40% additional sampling necessary for the sign test with a large number of samples. However, it may still be more economic to collect a lot of samples which can be quickly assessed than to collect fewer but much more time-consuming data. A nonparametric test will then be perfectly adequate for demonstrating the superiority of one treatment over another – but if you want to know how much more superior it is, then only a parametric test will do.

The bottom line is that it is better to use a nonparametric test and lose discriminatory power than to wrongly apply a parametric test.

Some ways data are organized for nonparametric tests

I've already warned you that I'm not going to equip you to carry out non-parametric tests to completion, but I do propose to show you how simple the initial stages of such tests are compared with parametric ones. I'll take a few nonparametric tests to the point where you'll need to move to a more advanced textbook. There you will find what are usually highly algebraic recipes for calculating the statistic often unique to that particular test. Equally unique are the tables required to assess the significance of that statistic.

The sign test

To illustrate this test, I'll repeat some data (Box 8.8) that we have already used for the paired t-test. They concern comparing two soils for how well ten plant species (effectively the replicates) grow in them when in containers. The data are dry weight in grams of the aerial parts of the plant when ready for sale:

Plant species	Soil A	Soil B
1	5.8	5.7
2	12.4	11.9
3	1.4	1.6
4	3.9	3.8
5	24.4	24.0
6	16.4	15.8
7	9.2	9.0
8	10.4	10.2
9	6.5	6.4
10	0.3	0.3
Mean	**9.07**	**8.87**

All we do is give each row of the table the + or − sign for column A − column B:

Plant species	Soil A	Soil B	
1	5.8	5.7	+
2	12.4	11.9	+
3	1.4	1.6	−
4	3.9	3.8	+
5	24.4	24.0	+
6	16.4	15.8	+
7	9.2	9.0	+
8	10.4	10.2	+
9	6.5	6.4	+
10	0.3	0.3	0
Mean	**9.07**	**8.87**	

The test statistic for the sign test is so simple I can actually tell you what it is here. It is the smaller of the number of plus or minus signs, which in the above example is the minus signs with just one! The test is not dissimilar from evaluating the probability of tossing a coin several times and, in this instance, getting just one head in 10 goes. In this example, the test statistic of 1 with $n = 10$ is just significant ($P = 0.05$).

The Kruskal–Wallis analysis of ranks

As soon as we have more than two treatments, we need a nonparametric test analogous to the analysis of variance, and the Kruskal–Wallis test is one of these. To illustrate how the data are manipulated, I'll again use a set of data we've used before; this time the data for the fully randomized design for analysis of variance (from page 118), which are days to flowering of 50% of broad bean plants per plot given one of three fertilizers, A, B, or C (with four replicates):

A	B	C
35	43	29
38	38	31
42	40	30
40	39	33
Mean	**Mean**	**Mean**
38.75	**40.00**	**30.75**

The first thing we do is to rank all 12 measurements regardless of their fertilizer treatment, with ties receiving the sum of their ranks divided by the number sharing that rank, i.e. if three plots were tied after position 8, they would be given the positions 9, 10, and 11 and the rank $(9 + 10 + 11)/3 = 10$.

The 12 plots are therefore arranged from lowest to highest as follows, with their rank in bold italics beneath:

29	30	31	33	35	38	38	39	40	40	42	43
1	*2*	*3*	*4*	*5*	*6*	*7*	*8*	*9*	*10*	*11*	*12*
					6.5	*6.5*		*9.5*	*9.5*		

Then we reproduce the table of data, but with each datum replaced by its rank from the above line-up:

A	B	C
5	*12*	*1*
6.5	*6.5*	*3*
11	*9.5*	*2*
9.5	*8*	*4*

Totalling the columns gives: **32.00   36.00   10.00**
whereas the data means were: 38.75   40.00   30.75

The test statistic (H) is based on the summed squares of the rank totals and the total number of observations (Box 19.1). Box 19.1 shows that, in this example, $H = 7.585$ ($P < 0.01$ from appropriate tables). The nonparametric test has therefore shown, like the analysis of variance on page 120, that we can reject the null hypothesis that data in the three columns were sampled from the same population.

This test can be used even when there are unequal numbers of data per treatment.

Kendall's rank correlation coefficient

As a final example of how data are handled for nonparametric tests, I'll use Kendall's rank correlation method to illustrate the form in which correlation data can be presented. I'll use the data on weight and length of broad bean seeds from the "spare-time" activities of Chapter 17 (page 276).

The first step is to arrange either the weight or length data in increasing order of magnitude, with the corresponding other measurement underneath (it doesn't matter which you choose to order, I'll use length). Where there are tied data, I arrange them in increasing order of the other measurement:

Length (cm)	1.6	1.7	1.9	2.0	2.0	2.1	2.2	2.2	2.3	2.4
Weight (g)	0.8	0.7	0.9	0.9	1.0	1.0	1.1	1.2	1.4	1.2

Next, taking the lower row (weight) we calculate three things for each number there, with each calculation adding a new row.

1 *The row "More"* (often designated as P): How many values to the right in the lower row are **larger** than that number. Note we don't bother with the extreme right value – there are no numbers further to the right to compare it with!

2 *The row "Less"* (often designated as Q): How many values to the right in the lower row are **smaller** than that number. Again, we don't need to bother with the last number.

3 *The row "Difference"* $(P - Q)$: The value in the "Less" row subtracted from the value in the "More" row:

Length (cm)		1.6	1.7	1.9	2.0	2.0	2.1	2.2	2.2	2.3	2.4
Weight (g)		0.8	0.7	0.9	0.9	1.0	1.0	1.1	1.2	1.4	1.2
More (P)		8	8	6	6	4	4	3	1	0	
Less (Q)		1	0	0	0	0	0	0	0	1	
Difference ($P - Q$)		+7	+8	+6	+6	+4	+4	+3	+1	−1	

The test statistic is the total of the values in the last row (in our case $+38$). For 10 pairs ($n = 10$) this shows significant correlation ($P < 0.05$ from appropriate tables). The calculation of the correlation coefficient (r) in our example is shown in Box 19.1 The value of r is 0.691, rather lower than the 0.896 calculated parametrically (from r^2 in "Solutions" on page 333).

The main nonparametric methods that are available

Finally in this chapter, which has introduced you to the idea of nonparametric methods, I will look at the various nonparametric tests you might wish to look up in more advanced statistical texts. I would guess that the most helpful way to do this is to list them by the kind of data for which you may find them useful and doing this by analogy with the parametric tests covered in the earlier chapters of this book. Remember, however, that nonparametric methods enable you to analyze data you could never hope to analyze by parametric methods (e.g. distribution-free data such as ranks, see earlier).

By using the analogy with parametric methods I am not implying that particular parametric and nonparametric tests are necessarily alternatives. It is merely the way the results table looks (not the type of data within it) that makes an analogy possible. In what follows, to keep things simple and comprehensible, I will use the word "treatments" in the broadest sense to identify a replicate group of data sharing the same "experimental" characteristic, e.g. male students, all plots treated with fertilizer A.

Analysis of two replicated treatments as in the t-test (Chapter 8)

For unpaired data of replicate values for the two treatments
   Example: Box 8.4 – comparing the statistic examination marks of 12 replicate and 14 replicate women.
   Nonparametric tests: *Mann–Whitney test* or *Kolmorgorov–Smirnov two-sample test.* Both tests can accommodate unequal or equal replicates per treatment, though the evaluation of statistical significance in the Kolmogorov–Smirnov test becomes more complicated with unequal replicate numbers. The Mann–Whitney test is more powerful for larger numbers of replicates, while the Kolmogorov–Smirnov test is preferred for a very small number of replicates.

For paired data of replicate values for the two treatments
   Example: Box 8.8 – growing ten different plants as replicates in two soils.
   Nonparametric tests: *Sign test* (see earlier) or *Wilcoxon's signed rank test.* The sign test is simpler, but merely works on the direction of differences, ignoring their magnitude. By contrast, the Wilcoxon test is more powerful in the sense that it gives greater weight to pairs showing a larger difference than to pairs showing only a small difference.

Analysis of more than two replicated treatments as in the analysis of variance (Chapter 11)

For fully randomized replicates of the treatments
   Example: page 117 – Days to flowering of broad beans treated with three fertilizers as a fully randomized design of four replicates per treatment.
   Nonparametric test: *Kruskal–Wallis analysis of ranks* (see earlier). Like the analysis of variance for a fully-randomized design, the Kruskal–Wallis test can accommodate unequal numbers of replicates in different treatments.

For experimental designs where the replicates of the treatments are grouped as Blocks. Nonparametric tests are limited to a two-way analysis – i.e. "Treatments" and Blocks. In factorial experiments (Chapters 13 and 14), therefore, each different factorial combination has to be regarded as a separate "treatment."

Example: page 123 – Days to flowering of broad beans treated with three fertilizers laid out in four replicate Blocks.

Nonparametric test: *Friedman two-way analysis.* This test, unlike the corresponding analysis of variance, cannot accommodate missing plots – numbers of replicates have to equal for all treatments.

Correlation of two variables (Chapter 17)

Example: Box 17.2 – actual and guessed number of mustard seeds.

Nonparametric tests: *Kendall's rank correlation coefficient* (see earlier) or *Spearman's rank correlation coefficient.* Both tests are equally powerful, but an advantage of Kendall's test is that it is possible to develop it to a correlation of one variable with **two** others.

Appendix 1
How many replicates

Introduction

When I have been acting as external examiner for PhD theses, I have often asked the candidate how the number of replicates for an experiment in the thesis had been chosen. The answer was either "it was what my supervisor suggested," or "I couldn't fit more into the glasshouse space I was allocated," or "I couldn't handle more in one day."

None of these are really good reasons – whatever the practical constraints may have been. If the number of replicates is too small (in relation to the variability of the material) to enable detection of differences that exist between treatments, then even "the limited glasshouse space" and "the days' work" have been totally wasted. Not doing the experiment at all would leave the experimenter equally informed, but with the research kitty still intact. At the other end of the scale, if the number of replicates is excessive, then scarce research resources of space and money have been squandered.

I don't think I ever got the answer "I worked out the optimum number of replicates statistically." Perhaps this is not surprising – I have found that how to do this is rather hidden away in many statistics textbooks and the topic has not even been mentioned in some statistics courses I have attended. So this Appendix is not just a casual add-on to this book; it discusses a subject of considerable practical importance which deserves more attention from biological researchers.

Underlying concepts

You need to be sure you have understood "standard error of differences between means" (page 52) and the t-test (page 59) before going any further.

The standard error of differences between two means (s.e.d.m.) is calculated from the variance of individual observations and the number of those observations (n) as:

$$\text{s.e.d.m.} = \sqrt{2 \times \text{variance}/n}$$

The s.e.d.m. will therefore get smaller as the divisor "n" increases. This has two important consequences which help us to design the optimum size of experiments:

1 If we have just one difference between means to test (as at the end of an experiment with two treatments), its chances of reaching statistical significance (i.e. exceeding $t \times$ s.e.d.m.) clearly increase as the s.e.d.m. shrinks, which can be achieved by increasing "n." This is analogous to trying to support a plank on two chairs further apart than the plank is long. You can't make the plank longer, so – if you want to succeed – you have to shrink the space between the chairs by moving them closer together. More replicates don't change the true difference between two means (i.e. the length of the plank), but they do reduce the chances of an experimental difference of a given size being "too short a plank" (Fig. A1.1) to attain significance.

2 If we have lots of sampled differences between pairs of means sampled at random from the **same population**, the differences would form a normal distribution about the true average difference, which is of course

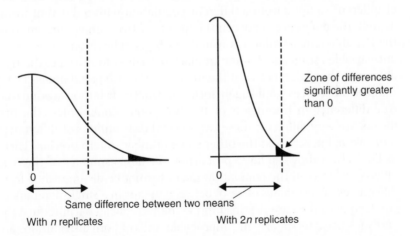

Fig. A1.1 Doubling the number of replicates in an experiment increases the probability that the **true difference** between the means of two treatments will fall into the "improbable" area of the distribution of differences between means drawn from the same population.

(a)   (b)

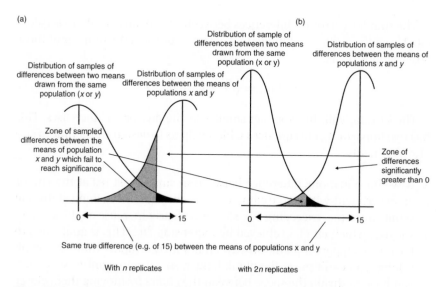

Fig. A1.2 Doubling the number of replicates in an experiment increases the probability that a higher **proportion of differences which might be sampled** between the means of two treatments (with a true difference of 15) will fall into the "improbable" area of the distribution of differences between means drawn from the same population.

zero in a single population (page 59). By contrast, if we now took lots of different samples from a different population with a different mean as well, the differences between the means of two samples drawn from the two different populations would not be variable estimates of zero, but variable estimates of some number not equal to zero, i.e. the true difference that exists between the means of the two populations. These sampled differences would again form a normal distribution around that true difference in the means of the two populations. Unless the two means are very different, these two normal distributions of differences (one averaging zero and the other some number) might overlap. Then, as Fig. A1.2a shows, a large proportion of these estimates of the true difference between the population mean cannot be distinguished from differences (all estimates of zero) between means in a single population. Analogous to the case desribed in (1), more replicates will reduce the yardstick (s.e.d.m.) by which "improbable" tails of both distributions are defined, so that more samples of differences between the means of the **two** populations will fall into the "improbable" tail of the distribution of differences from the **same** population (Fig. A1.2b). It's a bit like two wide triangles, one behind the other, positioned so that you can't see

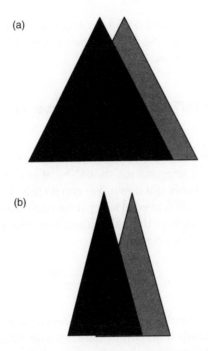

Fig. A1.3 The concept of Fig. A1.2 explained by analogy with two overlapping triangles, and the increase in area of the back triangle revealed when the bases of the triangle are narrowed without any change in the position of their peaks.

very much of the one at the back (Fig. A1.3a). As soon as the bases of the triangles are made narrower (Fig. A1.3b) a much greater proportion of the area of the triangle at the back separates from the one at the front, even though the positions of the peaks remain stationary! An over-simplified analogy, perhaps, but it may help?

Note: The above hopefully enables you to understand an important aspect of how increasing replication improves the chances of separating two means statistically. It is an important part of the story, but only part. There are two other effects that increasing the number of replicates has. Firstly, the size of t (by which the s.e.d.m. is multiplied to calculate the "least significant difference") decreases with more degrees of freedom (n − 1) as well as the s.e.d.m. does. This further helps the chances of statistically separating two means Secondly, with more replicates, our estimate of the difference between the two means gets nearer the true value. If our estimate from few replicates was inflated, it will get smaller with more replicates and therefore work against detecting the two means as significantly different. However, if we have underestimated the true difference between the means with our few replicates, it will get larger with more replicates (at the same time as a decreasing s.e.d.m.), and the chances of separating the means statistically obviously improve.

"Cheap and cheerful" calculation of number of replicates needed

This approach is based on Fig. A1.1. The calculation of the LSD as $t \times \sqrt{2 \times \text{variance}/n}$ suggests we should be able to solve for the "n" if we can make this the only unknown. We can do this if we put a numerical value on that LSD which is needed to separate two means statistically.

If we square everything, the calculation becomes $\text{LSD}^2 = t^2 \times 2 \times \text{variance}/n$ from which we can solve for:

$$n = \frac{t^2 \times 2 \times \text{variance}}{\text{LSD}^2.}$$

So in terms of the number of replicates needed for a difference between means to be significantly different (if there is a true difference to be found), the calculation becomes:

$$\text{Number of replicates} = \frac{t^2 \times 2 \times \text{variance}}{\text{difference}^2}$$

So to an example: Say we have an experiment with five treatments and four replicates. The residual mean square (= variance) is 0.48 ($4 \times 3 = 12$ d.f.) and we have not been able to detect that a difference between two means of 0.7 is statisitically significant at $P = 0.05$. How many replicates might have shown a statistically significant difference if one really exists? t for 12 d.f. at $P = 0.05$ is 2.179.

$$\text{Number of replicates needed} = \frac{4.75 \times 2 \times 0.48}{0.49} = 9.31$$

Ten replicates should be enough. I call this the "cheap and cheerful" method because, of course, t in an experiment with ten replicates would be smaller than 2.179, making detection of the difference (if it exists) even more certain!

The method can also be used with the variance obtained by some preliminary sampling **before** rather than after an experiment is attempted, using the number of samples $- 1$ as the d.f. for t. However, it will pay to repeat the calculation with a d.f. based on that of the residual mean square of the intended experiment given the number of replicates suggested by the initial calculation.

More accurate calculation of number of replicates needed

This approach is based on Fig. A1.2. As with the "cheap and cheerful" method, this method can also be used for planning the size of an experiment following some preliminary sampling to get an estimate of the variance. The difference between the methods is that, in the more accurate method, we allow for the reality that we cannot know how close our eventually sampled difference between two means will be to the true difference between them. In the "cheap and cheerful" approach we worked out the number of replicates needed to detect a fixed difference between two means. In the more accurate method, we visualize a normal distribution of the potential differences between two means that samples might show and then we have to make a decision on what area of the distribution of potential differences we wish to attain significance should a difference between the means really exist. This is illustrated in Fig. A1.2, where in (a) the area under the curve of differences between populations x and y which would show as significant (unshaded area) is smaller than in (b) (which has twice as many replicates). Putting it another way, fewer of the mean x – mean y differences in (b) fall into the area (shaded) which is not distinguishable from a curve around a mean of zero (the average of differences between means drawn from a single population).

So, in addition to using t for $P = 0.05$, we need to specifiy with what certainty we wish to detect any real difference between the means that exists (e.g. 60%, 70%, 80%, or 90% certainty?). Thus, if we opt for 80% or 90%, we have to enlarge the two 2.5% probability tails (to make 5% for $P = 0.05$) in the distribution of differences between the means by respectively 20% and 10%, or respectively 0.4 and 0.2 in "P" terms. The t value for the calculation then becomes a combination of two t values: that for t at $P = 0.05$ plus that for t at the required level of certainty ($P = 0.4$ or 0.2 in the description above). So that you can identify it, this combination of t values will be shown in bold in brackets in the calculations that follow.

To give the calculation procedure for the more accurate method, let's begin with a variance of 4.5, five treatments and a difference between means of 2.5 to be detected as significant (if such a difference really exists; it may not, of course). We decide we want a 90% certainty of detection at $P = 0.05$, which will involve us in two values of t, for $P = 0.05$ and $P = 0.20$ (for 90% certainty, see above). We now have to guess a number of replicates, say four. This makes 12 (5 treatments -1×4 replicates -1) the residual d.f. for looking up t, which therefore is 2.179 and 1.356 for $P = 0.05$ and $P = 0.20$ respectively.

Beginning with this guess of four replicates, we make our first calculation of the number of replicates which might really be required:

$$\text{Number of replicates} = \frac{[\mathbf{t}]^2 \times 2 \times \text{variance}}{(\text{difference to be detected})^2}$$

It's really pretty much the same as the "cheap and cheerful" calculation except for the addition of the certainty factor. So we'll turn the above expression into the numbers for our example:

$$t_{(P=0.05,\ 12\ \text{d.f.})} \qquad t_{(P=0.20,\ 12\ \text{d.f.})} \qquad \text{variance}$$

$$n = \frac{[2.179 + 1.356]^2 \times 2 \times 4.50}{2.5^2} = \frac{12.496 \times 9.000}{6.250} = 17.994$$

difference to be detected

This suggests 18 replicates, so we repeat the calculation assuming 18 replicates. A repeat calculation is necessary because we have now changed the residual degrees of freedom for t from 12 to $5 - 1$ treatments $\times 18 - 1$ replicates, i.e. 68. Table A2.1 only gives t values for 50 and 120 d.f.; I would just boldly guess 1.995 for $P = 0.05$ and 1.295 for $P = 0.20$. The new calculation is therefore:

$$n = \frac{[1.995 + 1.295]^2 \times 2 \times 4.500}{2.5^2} = \frac{10.824 \times 9.000}{6.250} = 15.566$$

The new answer of 16 replicates does not differ greatly from the 18 replicates we used for it, and I would "stick" at this point. A larger difference would cause me to recalculate again with the new guess. If I decided to do it here, the d.f. would now be $(5 - 1) \times (16 - 1) = 60$. The new calculation still comes out at 16 replicates, and would be:

$$n = \frac{[2.000 + 1.296]^2 \times 2 \times 4.500}{2.5^2} = \frac{10.864 \times 9.000}{6.250} = 15.644$$

Sometimes the first answer is stupifyingly large, say 1889! Who could do that number of replicates? But doesn't it just show how useful the exercise is in alerting you that you might have been proposing a complete waste of time? Should this befall you, all is not lost:

- A ridiculously large number of replicates is often a sign that the variance is large compared with the difference to be detected. This is usually because the data are not normally distributed around the mean but badly clumped. If this turns out to be true, then transforming the data (e.g. to square root + 0.5) as discussed on page 38 can work miracles! The variance of the transformed values can shrink dramatically (of course the per cent difference to be detected will also need to be recalculated for the transformed data). In a student exercise involving the sampling of the very clumped whitefly population found on the shrub *Viburnum tinus*, we found that the 450 replicates needed to detect a doubling of the population shrunk to about 40 after applying the square root transformation.
- The second thing that will help is to lower our sights in relation to the size of difference we want to be sure we could detect. If we cannot detect a 25% difference with the maximum replication that is practicable, is there any value in doing the experiment if we can only detect a 50% difference? There may well be. In assessing varieties of beans for resistance to blackfly, it might be nice to identify ones that lower the aphid population by 25%, but that might not be enough reduction to satisfy gardeners. However, they might buy the seed of a variety known to halve the population.
- The temptation to go for maximum certainty of finding a significant difference between means is very tempting, so most people would probably do the first calculation with a nominated certainty of 90% or over. However, if it becomes a choice between abandonig the experiment because of the impractical replication needed or lowering the level of certainty, you should certainly consider taking a chance on the latter.

How to prove a negative

It may be possible to "disprove" the null hypothesis (page 59) to the extent of showing that the probability of two different mean values arising by chance sampling of a single population is less than 0.05. It is much harder to "prove" that the null hypothesis is true! Would the null hypothesis not have held if you had done a few more replicates? At least the above calculation does put you in the position of being able to say "If there **had** been a real difference, then I really should have found it."

Appendix 2
Statistical tables

Table A2.1 Values of the statistic t. (Modified from *Biometrika Tables for Statisticians*, vol. 1, eds E.S. Pearson and H.O. Hartley (1958), published by Cambridge University Press, by kind permission of the publishers and the Biometrika Trustees.)

d.f.	Probability of a larger values of t (two-tailed test)								
	0.5	0.4	0.3	0.2	0.1	0.05	0.02	0.01	0.001
1	1.000	1.376	1.963	3.078	6.314	12.71	31.82	63.66	636.62
2	.816	1.061	1.386	1.886	2.920	4.303	6.965	9.925	31.60
3	.765	.978	1.250	1.638	2.353	3.182	4.541	5.841	12.94
4	.741	.941	1.190	1.533	2.132	2.776	3.747	4.604	8.610
5	.727	.920	1.156	1.476	2.015	2.571	3.365	4.032	6.859
6	.718	.906	1.134	1.440	1.943	2.447	3.143	3.707	5.959
7	.711	.896	1.119	1.415	1.895	2.365	2.998	3.499	5.405
8	.706	.889	1.108	1.397	1.860	2.306	2.896	3.355	5.041
9	.703	.883	1.100	1.383	1.833	2.262	2.821	3.250	4.781
10	.700	.879	1.093	1.372	1.812	2.228	2.764	3.169	4.587
11	.697	.876	1.088	1.363	1.796	2.201	2.718	3.106	4.437
12	.695	.873	1.083	1.356	1.782	2.179	2.681	3.055	4.318
13	.694	.870	1.079	1.350	1.771	2.160	2.650	3.012	4.221
14	.692	.868	1.076	1.345	1.761	2.145	2.624	2.977	4.140
15	.691	.866	1.074	1.341	1.753	2.131	2.602	2.947	4.073
16	.690	.865	1.071	1.337	1.746	2.120	2.583	2.921	4.015
17	.689	.863	1.069	1.333	1.740	2.110	2.567	2.898	3.965
18	.688	.862	1.067	1.330	1.734	2.101	2.552	2.878	3.922
19	.688	.861	1.066	1.328	1.729	2.093	2.539	2.861	3.883
20	.687	.860	1.064	1.325	1.725	2.086	2.528	2.845	3.850
21	.686	.859	1.063	1.323	1.721	2.080	2.518	2.831	3.819
22	.686	.858	1.061	1.321	1.717	2.074	2.508	2.819	3.792
23	.685	.858	1.060	1.319	1.714	2.069	2.500	2.807	3.767
24	.685	.857	1.059	1.318	1.711	2.064	2.492	2.797	3.745
25	.684	.856	1.058	1.316	1.708	2.060	2.485	2.787	3.725
26	.684	.856	1.058	1.315	1.706	2.056	2.479	2.779	3.707
27	.684	.855	1.057	1.314	1.703	2.052	2.473	2.771	3.690
28	.683	.855	1.056	1.313	1.701	2.048	2.467	2.763	3.674
29	.683	.854	1.055	1.311	1.699	2.045	2.462	2.756	3.659
30	.683	.854	1.055	1.310	1.697	2.042	2.457	2.750	3.646
40	.681	.851	1.050	1.303	1.684	2.021	2.423	2.704	3.551
60	.679	.848	1.046	1.296	1.671	2.000	2.390	2.660	3.460
120	.677	.845	1.041	1.289	1.658	1.980	2.358	2.617	3.373
∞	.674	.842	1.036	1.282	1.645	1.960	2.326	2.576	3.291

Table A2.2 Multipliers for the multiple range test.

d.f.	Number of treatments																		
	2	3	4	5	6	7	8	9	10	11	12	13	14	15	16	17	18	19	20
1	12.7	19.1	23.2	26.3	28.6	30.5	32.1	33.4	34.7	35.8	36.7	37.6	38.4	39.2	39.8	40.4	41.0	41.6	42.1
2	4.31	5.89	6.93	7.70	8.29	8.79	9.21	9.57	9.89	10.18	10.43	10.66	10.88	11.07	11.25	11.41	11.57	11.72	11.86
3	3.18	4.18	4.83	5.31	5.69	5.99	6.26	6.49	6.69	6.87	7.04	7.18	7.32	7.44	7.56	7.67	7.76	7.86	7.95
4	2.78	3.56	4.07	4.45	4.74	4.99	5.20	5.37	5.54	5.68	5.81	5.92	6.06	6.13	6.22	6.31	6.39	6.46	6.53
5	2.57	3.25	3.69	4.01	4.26	4.48	4.65	4.81	4.94	5.07	5.18	5.28	5.37	5.46	5.54	5.61	5.68	5.74	5.81
6	2.45	3.07	3.46	3.75	3.98	4.16	4.33	4.47	4.59	4.70	4.80	4.89	4.98	5.05	5.12	5.19	5.25	5.31	5.37
7	2.36	2.94	3.31	3.58	3.78	3.95	4.10	4.24	4.35	4.45	4.54	4.62	4.70	4.77	4.84	4.90	4.96	5.01	5.06
8	2.31	2.86	3.20	3.46	3.66	3.82	3.96	4.08	4.19	4.28	4.37	4.45	4.52	4.58	4.65	4.70	4.76	4.81	4.86
9	2.26	2.79	3.13	3.37	3.55	3.71	3.84	3.96	4.06	4.15	4.23	4.31	4.38	4.44	4.50	4.55	4.60	4.65	4.70
10	2.23	2.74	3.06	3.30	3.47	3.62	3.75	3.86	3.96	4.04	4.12	4.19	4.26	4.33	4.38	4.43	4.48	4.53	4.57
11	2.20	2.70	3.01	3.24	3.41	3.56	3.68	3.78	3.88	3.97	4.04	4.11	4.17	4.23	4.29	4.34	4.38	4.43	4.48
12	2.18	2.67	2.97	3.19	3.36	3.50	3.62	3.73	3.82	3.90	3.97	4.04	4.10	4.16	4.21	4.26	4.31	4.35	4.39
13	2.16	2.64	2.93	3.15	3.32	3.45	3.57	3.67	3.76	3.84	3.91	3.98	4.04	4.09	4.14	4.19	4.24	4.29	4.32
14	2.14	2.62	2.91	3.12	3.28	3.42	3.53	3.63	3.71	3.79	3.86	3.93	3.99	4.04	4.09	4.14	4.19	4.23	4.26
15	2.13	2.60	2.88	3.09	3.25	3.38	3.49	3.59	3.68	3.75	3.82	3.88	3.94	4.00	4.04	4.09	4.14	4.18	4.21
16	2.12	2.58	2.86	3.07	3.22	3.35	3.46	3.56	3.64	3.71	3.78	3.85	3.90	3.95	4.00	4.05	4.09	4.13	4.17
17	2.11	2.56	2.84	3.05	3.20	3.32	3.44	3.53	3.61	3.68	3.75	3.81	3.87	3.92	3.97	4.02	4.06	4.09	4.13
18	2.10	2.55	2.83	3.03	3.17	3.30	3.42	3.51	3.59	3.66	3.73	3.78	3.84	3.89	3.94	3.98	4.02	4.06	4.09
19	2.09	2.54	2.81	3.01	3.16	3.28	3.39	3.48	3.56	3.63	3.70	3.76	3.81	3.86	3.91	3.95	4.00	4.03	4.07
20	2.09	2.53	2.80	3.00	3.15	3.27	3.37	3.46	3.54	3.61	3.68	3.73	3.79	3.84	3.89	3.93	3.97	4.00	4.04
24	2.06	2.50	2.76	2.95	3.09	3.21	3.31	3.40	3.48	3.54	3.61	3.66	3.71	3.76	3.80	3.85	3.89	3.92	3.95
30	2.04	2.46	2.72	2.91	3.04	3.15	3.25	3.34	3.42	3.48	3.54	3.59	3.64	3.68	3.73	3.77	3.80	3.87	3.87
40	2.02	2.43	2.68	2.86	2.99	3.10	3.20	3.27	3.35	3.42	3.46	3.52	3.57	3.61	3.66	3.69	3.73	3.76	3.79
60	2.00	2.40	2.64	2.81	2.94	3.05	3.14	3.22	3.29	3.34	3.40	3.45	3.49	3.54	3.58	3.61	3.64	3.68	3.71
120	1.98	2.38	2.61	2.77	2.90	3.00	3.08	3.16	3.22	3.28	3.33	3.38	3.42	3.46	3.50	3.54	3.56	3.60	3.63
∞	1.96	2.35	2.57	2.73	2.85	2.95	3.03	3.10	3.16	3.22	3.27	3.31	3.35	3.39	3.42	3.46	3.49	3.51	3.54

Table A2.3 Values of the statistic F. (Modified from *Biometrika Tables for Statisticians*, vol. 1, eds E.S. Pearson and H.O. Hartley (1958), published by Cambridge University Press, by kind permission of the publishers and the Biometrika Trustees.)

The three rows for each degree of freedom are the maximum values of F for respectively $P = 0.05$, $P = 0.01$ and $P = 001$ (one-tailed test)

d.f. for smaller Mean Square	d.f. for larger Mean Square									
	1	2	3	4	5	6	8	12	24	∞
2	18.51	19.00	19.16	19.25	19.30	19.33	19.37	19.41	19.45	19.50
	98.49	99.01	99.17	99.25	99.30	99.33	99.36	99.42	99.46	99.50
	998	999	999	999	999	999	999	999	999	999
3	10.13	9.55	9.28	9.12	9.01	8.94	8.84	8.74	8.64	8.53
	34.12	30.81	29.46	28.71	28.24	27.91	27.49	27.05	26.60	26.12
	167.0	148.5	141.1	137.1	134.6	132.8	130.6	128.3	125.9	123.5
4	7.71	6.94	6.59	6.39	6.26	6.16	6.04	5.91	5.77	5.63
	21.20	18.00	16.69	15.98	15.52	15.21	14.80	14.37	13.93	13.46
	74.13	61.24	56.18	53.48	51.71	50.52	49.00	47.41	45.77	44.05
5	6.61	5.79	5.41	5.19	5.05	4.95	4.82	4.68	4.53	4.36
	16.26	13.27	12.86	11.39	10.97	10.67	10.27	9.89	9.47	9.02
	47.04	36.61	33.28	31.09	29.75	28.84	27.64	26.41	25.15	23.78
6	5.99	5.14	4.76	4.53	4.39	4.28	4.15	4.00	3.84	3.67
	13.74	10.92	9.78	9.15	8.75	8.47	8.18	7.72	7.31	6.88
	35.51	27.00	23.70	21.90	20.81	20.03	19.03	17.99	16.89	15.75
7	5.59	4.74	4.35	4.12	3.97	3.87	3.73	3.57	3.41	3.23
	12.25	9.55	8.45	7.85	7.46	7.19	6.84	6.47	6.07	5.65
	29.25	21.69	18.77	17.20	16.21	15.52	14.63	13.71	12.73	11.70
8	5.32	4.46	4.07	3.84	3.69	3.58	3.44	3.28	3.12	2.93
	11.26	8.65	7.59	7.01	6.63	6.37	6.03	5.67	5.23	4.86
	25.42	18.49	15.83	14.39	13.49	12.86	12.05	11.20	10.30	9.34
9	5.12	4.26	3.86	3.63	3.48	3.37	3.23	3.07	2.90	2.71
	10.56	8.02	6.99	6.42	6.06	5.80	5.47	5.11	4.73	4.31
	22.86	16.59	13.98	12.50	11.71	11.13	10.37	9.57	8.72	7.81
10	4.96	4.10	3.71	3.48	3.33	3.22	3.07	2.91	2.74	2.54
	10.04	7.56	6.55	5.99	5.64	5.39	5.06	4.71	4.33	3.91
	21.04	14.91	12.55	11.28	10.48	9.93	9.20	8.44	7.64	6.76
11	4.64	3.98	3.59	3.36	3.28	3.09	2.95	2.79	2.61	2.40
	9.65	7.20	6.22	5.67	5.32	5.07	4.74	4.40	4.02	3.60
	19.68	13.81	11.56	10.35	9.58	9.05	8.35	7.63	6.85	6.00
12	4.75	3.38	3.49	3.26	3.11	3.00	2.85	2.69	2.50	2.30
	9.33	6.93	5.95	5.41	5.06	4.82	4.50	4.16	3.78	3.36
	18.64	12.97	10.81	9.63	6.89	8.38	7.71	7.00	6.25	5.42

Table A2.3   Continued

d.f. for smaller Mean Square	d.f. for larger Mean Square									
	1	2	3	4	5	6	8	12	24	∞
13	4.67	3.80	3.41	3.18	3.02	2.92	2.77	2.60	2.42	2.21
	9.07	6.70	5.74	5.20	4.86	4.62	4.30	3.96	3.59	3.16
	17.81	12.31	10.21	9.07	8.35	7.86	7.21	6.52	5.78	4.97
14	4.60	3.74	3.34	3.11	2.96	2.85	2.70	2.53	2.35	2.13
	8.86	6.51	5.56	5.03	4.69	4.46	4.14	3.80	3.43	3.00
	17.14	11.78	9.73	8.62	7.92	7.43	6.80	6.13	5.41	4.60
15	4.54	3.68	3.29	3.00	2.90	2.79	2.64	2.48	2.29	2.07
	8.68	6.36	5.42	4.89	4.56	4.32	4.00	3.67	3.29	2.87
	16.58	11.34	9.34	8.25	7.57	7.09	6.47	5.81	5.10	4.31
16	4.49	3.63	3.234	3.01	2.85	2.74	2.59	2.42	2.24	2.01
	8.53	6.23	5.29	4.77	4.44	4.20	3.89	3.55	3.18	2.75
	16.12	10.97	9.00	7.94	7.27	6.80	6.20	5.55	4.85	4.06
17	4.45	3.59	3.20	2.96	2.81	2.70	2.55	2.38	2.19	1.96
	8.40	6.11	5.18	4.67	4.34	4.10	3.79	3.45	3.08	2.65
	15.72	10.66	8.73	7.68	7.02	6.56	5.96	5.32	4.63	3.85
18	4.41	3.55	3.16	2.93	2.77	2.66	2.51	2.34	2.15	1.92
	8.28	6.01	5.09	4.58	4.25	4.01	3.71	3.37	3.01	2.57
	15.38	10.39	8.49	7.46	6.81	6.35	5.76	5.13	4.45	3.67
19	4.38	3.52	3.13	2.90	2.74	2.63	2.48	2.31	2.11	1.83
	8.18	5.93	5.01	4.50	4.17	3.94	3.63	3.30	2.92	2.49
	15.08	10.16	8.28	7.27	6.61	6.18	5.59	4.97	4.29	3.51
20	4.35	3.49	3.10	2.87	2.71	2.60	2.45	2.28	2.08	1.84
	8.10	5.85	4.94	4.43	4.10	3.87	3.56	3.23	2.86	2.42
	14.82	9.95	8.10	7.10	6.46	6.02	5.44	4.82	4.15	3.38
21	4.32	3.47	3.07	2.84	2.68	2.57	2.42	2.25	2.05	1.81
	8.02	5.78	4.87	4.37	4.04	3.81	3.51	3.17	2.80	2.36
	14.59	9.77	7.94	6.95	6.32	5.88	5.31	4.70	4.03	3.26
22	4.30	3.44	3.05	2.82	2.66	2.55	2.40	2.23	2.03	1.78
	7.94	5.72	4.82	4.31	3.99	3.75	3.45	3.12	2.75	2.30
	14.38	9.61	7.80	6.81	6.19	5.76	5.19	4.58	3.92	3.15
23	4.28	3.42	3.03	2.80	2.64	2.53	2.38	2.20	2.00	1.76
	7.88	5.66	4.76	4.26	3.94	3.71	3.41	3.07	2.70	2.26
	14.19	9.47	7.67	6.70	6.08	5.65	5.09	4.48	3.82	3.05
24	4.26	3.40	3.01	2.78	2.62	2.51	2.36	2.18	1.98	1.73
	7.82	5.61	4.72	4.22	3.90	3.67	3.36	3.03	2.66	2.21
	14.03	9.34	7.55	6.59	5.98	5.55	4.99	4.39	3.74	2.97
25	4.24	3.38	2.99	2.76	2.60	2.49	2.34	2.16	1.96	1.71
	7.77	5.57	4.68	4.18	3.86	3.63	3.32	2.99	2.62	2.17
	13.87	9.22	7.45	6.49	5.89	5.46	4.91	4.31	3.66	2.89

(Continued)

Table A2.3   Continued

d.f. for smaller Mean Square	d.f. for larger Mean Square									
	1	2	3	4	5	6	8	12	24	∞
26	4.22	3.37	2.98	2.74	2.59	2.47	2.32	2.15	1.95	1.69
	7.72	5.53	4.64	4.14	3.82	3.59	3.29	2.96	2.58	2.13
	13.74	9.12	7.36	6.41	5.80	5.38	4.83	4.24	3.59	2.82
27	4.21	3.35	2.96	2.73	2.57	2.46	2.30	2.13	1.93	1.67
	7.68	5.49	4.60	4.11	3.78	2.56	3.26	2.93	2.55	2.10
	13.61	9.02	7.27	6.33	5.73	5.31	4.76	4.17	3.52	2.75
28	4.20	3.34	2.95	2.71	2.56	2.44	2.29	2.12	1.91	1.65
	7.64	5.45	4.57	4.07	3.75	3.53	3.23	2.90	2.52	2.06
	13.50	8.93	7.19	6.25	5.66	5.24	4.69	4.11	3.46	2.69
29	4.18	3.33	2.93	2.70	2.54	2.43	2.28	2.10	1.90	1.64
	7.60	5.42	4.54	4.04	3.73	3.50	3.20	2.87	2.49	2.03
	13.39	8.85	7.12	6.19	5.59	5.18	4.65	4.05	3.41	2.64
30	4.17	3.32	2.92	2.69	2.53	2.42	2.27	2.09	1.89	1.62
	7.56	5.39	4.51	4.02	3.70	3.47	3.17	2.84	2.47	2.01
	13.29	8.77	7.05	6.12	5.53	5.12	4.58	4.00	3.36	2.59
35	4.12	3.26	2.87	2.64	2.48	2.37	2.22	2.04	1.83	1.57
	7.42	5.27	4.40	3.91	3.59	3.37	3.07	2.74	2.37	1.90
	12.89	8.47	6.79	5.88	5.30	4.89	4.36	3.79	3.16	2.38
40	4.08	3.23	2.84	2.61	2.45	2.34	2.18	2.00	1.79	1.52
	7.31	5.18	4.31	3.83	2.51	3.29	2.99	2.66	2.29	1.82
	12.61	8.25	6.60	5.70	5.13	4.73	4.21	3.64	3.01	2.23
45	4.06	3.21	2.81	2.58	2.42	2.31	2.15	1.97	1.76	1.48
	7.23	5.11	4.25	3.77	3.45	3.23	2.94	2.61	2.23	1.75
	12.39	8.09	6.45	5.56	5.00	4.61	4.09	3.53	2.90	2.12
50	4.03	3.18	2.79	2.56	2.40	2.29	2.13	1.95	1.74	1.44
	7.17	5.06	4.20	3.72	3.41	3.19	2.89	2.56	2.18	1.68
	12.22	7.96	6.34	5.46	4.90	4.51	4.00	3.44	2.82	2.02
60	4.00	3.15	2.76	2.52	2.37	2.25	2.10	1.92	1.70	1.39
	7.08	4.98	4.13	3.65	3.34	3.12	2.82	2.50	2.12	1.60
	11.97	7.77	6.17	5.31	4.76	4.37	3.87	3.31	2.69	1.90
70	3.98	3.13	2.74	2.50	2.35	2.23	2.07	1.89	1.67	1.35
	7.01	4.92	4.07	3.60	3.29	3.07	2.78	2.45	2.07	1.53
	11.80	7.64	6.06	5.20	4.66	4.28	3.77	3.23	2.61	1.78
80	3.96	3.11	2.72	2.49	2.33	2.21	2.06	1.88	1.65	1.31
	6.96	4.88	4.04	3.56	3.26	3.04	2.74	2.42	2.03	1.47
	11.67	7.54	5.97	5.12	4.58	4.20	3.70	3.16	2.54	1.72
90	3.95	3.10	2.71	2.17	2.32	2.20	2.04	1.86	1.64	1.28
	6.92	4.85	4.01	3.53	3.23	3.01	2.72	2.39	2.00	1.43
	11.57	7.47	5.91	5.06	4.53	4.15	3.87	3.11	2.50	1.66

Table A2.3 Continued

d.f. for smaller Mean Square	d.f. for larger Mean Square								
	1	2	3	4	5	6	8	12	24	∞
100	3.94	3.09	2.70	2.46	2.30	2.19	2.03	1.85	1.63	1.26
	6.90	4.82	3.98	3.51	3.21	2.99	2.69	2.37	1.98	1.39
	11.50	7.41	5.86	5.02	4.48	4.11	3.61	3.07	2.46	1.61
∞	3.84	3.00	2.60	2.37	2.21	2.09	1.94	1.75	1.52	1.00
	6.64	4.62	3.78	3.32	3.02	2.80	2.51	2.18	1.79	1.00
	10.83	6.91	5.42	4.62	4.10	3.74	3.27	2.74	2.13	1.00

Table A2.4 Values of the statistic F_{max}. (From *Biometrika Tables for Statisticians*, vol. 1, eds E.S. Pearson and H.O. Hartley (1958), published by Cambridge University Press, by kind permission of the publishers and the Biometrika Trustees.)

Divide the largest of several variances by the smallest to obtain the statistic F_{max}. If this value is smaller than the value in the table (where the n of $n-1$ is the number of replicates of each treatment), the variances can be assumed to be homogenous.

$n-1$	Number of treatments										
	2	3	4	5	6	7	8	9	10	11	12
2	39.0	87.5	142	202	266	333	403	475	550	626	704
3	15.4	27.8	39.2	50.7	62.0	72.9	83.5	93.9	104	114	124
4	9.60	15.5	20.6	25.2	29.5	33.6	37.5	41.1	44.6	48.0	51.4
5	7.15	10.8	13.7	16.3	18.7	20.8	22.9	24.7	26.5	28.2	29.9
6	5.82	8.38	10.4	12.1	13.7	15.0	16.3	17.5	18.6	19.7	20.7
7	4.99	6.94	8.44	9.70	10.8	11.8	12.7	13.5	14.3	15.1	15.8
8	4.43	6.00	7.18	8.12	9.03	9.78	10.5	11.1	11.7	12.2	12.7
9	4.03	5.34	6.31	7.11	7.80	8.41	8.95	9.45	9.91	10.3	10.7
10	3.72	4.85	5.67	6.34	6.92	7.42	7.87	8.28	8.66	9.01	9.34
12	3.28	4.16	4.79	5.30	5.72	6.09	6.42	6.72	7.00	7.25	7.48
15	2.86	3.54	4.01	4.37	4.68	4.95	5.19	5.40	5.59	5.77	5.93
20	2.46	2.95	3.29	3.54	3.76	3.94	4.10	4.24	4.37	4.49	4.59
30	2.07	2.40	2.61	2.78	2.91	3.02	3.12	3.21	3.29	3.36	3.39
60	1.67	1.85	1.96	2.04	2.11	2.17	2.22	2.26	2.30	2.33	2.36
∞	1.00	1.00	1.00	1.00	1.00	1.00	1.00	1.00	1.00	1.00	1.00

Table A2.5 Values of the statistic χ^2. (Abridged with kind permission from NIST/SEMATECH e-Handbook of Statistical Methods – www.itl.nist.gov/div898/handbook, June 2003.)

d.f.	Probability of exceeding the critical value				
	0.10	0.05	0.025	0.01	0.001
1	2.706	3.841	5.024	6.635	10.828
2	4.605	5.991	7.378	9.210	13.816
3	6.251	7.815	9.348	11.345	16.266
4	7.779	9.488	11.143	13.277	18.467
5	9.236	11.070	12.833	15.086	20.515
6	10.645	12.592	14.449	16.812	22.458
7	12.017	14.067	16.013	18.475	24.322
8	13.362	15.507	17.535	20.090	26.125
9	14.684	16.919	19.023	21.666	27.877
10	15.987	18.307	20.483	23.209	29.588
11	17.275	19.675	21.920	24.725	31.264
12	18.549	21.026	23.337	26.217	32.910
13	19.812	22.362	24.736	27.688	34.528
14	21.064	23.685	26.119	29.141	36.123
15	22.307	24.996	27.488	30.578	37.697
16	23.542	26.296	28.845	32.000	39.252
17	24.769	27.587	30.191	33.409	40.790
18	25.989	28.869	31.526	34.805	42.312
19	27.204	30.144	32.852	36.191	43.820
20	28.412	31.410	34.170	37.566	45.315
21	29.615	32.671	35.479	38.932	46.797
22	30.813	33.924	36.781	40.289	48.268
23	32.007	35.172	38.076	41.638	49.728
24	33.196	36.415	39.364	42.980	51.179
25	34.382	37.652	40.646	44.314	52.620
26	35.563	38.885	41.923	45.642	54.052
27	36.741	40.113	43.195	46.963	55.476
28	37.916	41.337	44.461	48.278	56.892
29	39.087	42.557	45.722	49.588	58.301
30	40.256	43.773	46.979	50.892	59.703
35	46.059	49.802	53.203	57.342	66.619
40	51.805	55.758	59.342	63.691	73.402
45	57.505	61.656	65.410	69.957	80.077
50	63.167	67.505	71.420	76.154	86.661
60	74.397	79.082	83.298	88.379	99.607
70	85.527	90.531	95.023	100.425	112.317
80	96.578	101.879	106.629	112.329	124.839
90	107.565	113.145	118.136	124.116	137.208
100	118.498	124.342	129.561	135.807	149.449

Appendix 3
Solutions to "Spare-time activities"

The aim of these "solutions" is not only to provide the end answer, but also sufficient of the working towards that answer to enable you to identify – if you have gone wrong – exactly at what stage in the computation you have lacked the necessary understanding of the calculation involved.

However, pushing buttons on a calculator can be rather error prone. Don't be surprised if a wrong answer turns out **not** to be your failure to grasp the statistical procedures but merely "human error"!

Answers will also differ slightly depending on how many decimal places you keep when you write down intermediate answers from a calculator and use it for further calculations (look at Box A3.1 if you need convincing). So don't expect total agreement with my numbers, but agreement should be close enough to satisfy you that you have followed the correct statistical procedures.

Chapter 3

1 Here are 9 numbers totalling 90; the mean for calculating the deviations is therefore **10**.

The 9 respective deviations from the mean are therefore -1, 0, $+3$, -4, -2, $+2$, $+3$, 0, and -1. Squaring and adding these deviations should give the answer **44** (minus deviations become $+$ when squared).

The variance of the numbers is these summed squared deviations from the mean divided by the degrees of freedom $(n-1)$, i.e. $44/8 = \mathbf{5.50}$.

2 The total of the 16 numbers is 64; the mean is **4**.

Squaring and adding the 16 deviations from the mean (-3, -1, -2, $+2$, etc.) $= \mathbf{40}$.

Variance $= 40/(n-1) = 40/15 = \mathbf{2.67}$.
Standard deviation $= \sqrt{2.67} = \mathbf{1.63}$.
Mean $\pm$ standard deviation $= 4 \pm 1.63$.

BOX A3.1

On the left are different versions of the figure 5.555556 rounded (up in this case) to decreasing numbers of decimal places, and on the right the result of squaring that figure, with the answer rounded to the same number of decimal places:

5.555556	squared =	30.864202
5.55556	squared =	30.86425
5.5556	squared =	30.8645
5.556	squared =	30.869
5.56	squared =	30.91
5.6	squared =	31.4

This is the result of just one operation! With successive further mathematical steps then involving that number with others, additional variations in the digits will develop.

---

8 is a deviation of 4 from the mean. It is therefore $4/1.63 = \mathbf{2.45}$ standard deviations from the mean.

Chapter 4

1 (a) Correction factor $\left(\left(\sum x\right)^2 / n\right) = 90^2/9 = 900$.

Added squares $\sum x^2 = 9^2 + 10^2 + 13^2 + \cdots + 9^2 = 944$.

Sum of squares of deviations from the mean $\left(\sum x^2 - \left(\sum x\right)^2/n\right) = 944 - 900 = \mathbf{44}$.

Variance $= \mathbf{44/8} = \mathbf{2.67}$. Checks with solution for exercise in Chapter 3.

(b) Correction factor $\left(\left(\sum x\right)^2 / n\right) = 64^2/16 = 256$.

Added squares $\sum x^2 = 1^2 + 3^2 + 2^2 + \cdots + 2^2 = 296$.

Sum of squares of deviations from the mean $\left(\sum x^2 - \left(\sum x\right)^2/n\right) = 296 - 256 = \mathbf{40}$.

Variance $= \mathbf{40/15} = \mathbf{5.5}$. Checks with solution for exercise in Chapter 3.

2 Mean $= 2.47$. Variance $= 0.0041$. Standard deviation (s) $= 0.084$.

$2.43 = $ mean $- 0.045 = $ mean $- (0.045/0.064)s = \mathbf{2.47 - 0.703s}$.

$2.99 = $ mean $+ 0.515 = $ mean $+ (0.515/0.064)s = \mathbf{2.47 + 8.05s}$.

Chapter 7

The key to these calculations is to realize that the variance gives you s^2/n, which when doubled gives $2s^2/n$. The square root of this will be the standard error of differences between two means.

1 (a) Variance (of means of 10 plants) $= 1633$.

Therefore if $s^2/10 = 1633$, then s^2 (i.e. variance of individual plants) is $1633 \times 10 = \mathbf{16,330}$.

(b) Variance (of means of 10 plants) $= 1633$.

Therefore variance of differences between two means $= 3266$.

Therefore s.e.d.m. $= \mathbf{57.15}$.

The actual difference (**223.1 g** on one side and **240 g** on the other side of the path) is **16.9 g**, only **0.3 s.e.d.m. worths!**

We can therefore conclude that the two areas could be combined.

2 Variance (of means of 5 guesses per observer) $= 285.5$.

Therefore variance of differences between two means $= 571$.

Therefore s.e.d.m. $= \mathbf{23.9}$.

Thus any guess of $2 \times 23.9 = 47.8$ higher for any observer would safely suggest a larger number of seeds.

As the true mean of the first sample was 163, **any observer should be able to distinguish between 163 and $(163 + 47.8) = 211$ seeds in a mean of five guesses.**

3 (a) Standard deviation $= \sqrt{37.4} = 6.12$, so **mean $+$ 2 s** $= 135 + 12.24 = \mathbf{147.24\ g}$.

(b) Standard error of 5 leeks $= \sqrt{37.4/5}$ or $6.12/\sqrt{5} = 2.73$, so **mean $-$ 0.7 s.e.** $= 135 - 1.91 = \mathbf{133.09\ g}$.

(c) $1.4 \times 6.12 = \mathbf{8.57\ g}$.

(d) s.e.d.m. for means of 8 leeks $= \sqrt{2 \times 37.4/8} = 9.35$, therefore **a mean of 8 leeks of $135 - (2 \times 9.35) = 116.3\ g$ is the maximum weight significantly lower than 135 g.**

Chapter 8

1	Var. A		Var. B
Mean	4.100		4.484
Difference		0.384	
Sum of squares	5.540		3.565
Pooled variance		0.253	
s.e.d.m.			0.163
Calculated t		2.354	
t for $P = 0.05_{36\ \text{d.f.}}$		2.030 (35 d.f. is nearest lower value in table)	

Var. B significantly ($P < 0.05$) outyields Var. A by 0.384 kg per plot.

2

	Compost	Soil
Mean	42.67	28.90
Difference	13.77	
Sum of squares	318.00	92.90
Pooled variance	15.804	
s.e.d.m.		1.568
Calculated t	8.780	
t for $P = 0.05_{26d}$	2.056	

Even at $P < 0.001$, the grower can be satisfied that bought-in compost gives superior (about 52%) germination of *Tagetes* to his own sterilized soil.

3

	Constituent A	Constituent B
Mean	5.88	5.00
Difference	0.88	

Working on differences between A and B:

Correction factor	10.173 (keeping + and − signs)
Sum of squares	9.337
Variance	0.7814
Variance of means of	0.0601 ($= 0.7814/13$)
13 differences	
s.e.d.m.	0.2451
Calculated t	3.590
t for $P = 0.05_{12\ d.f.}$	2.179

The experimenter can be 99% sure ($P < 0.01$) that he has shown that constituent B does not run as far (mean of 13 runs is 0.88 cm less) as constituent A. It can therefore be deduced that the two constituents have different physicochemical properties, i.e. they are different compounds.

Chapter 11

1 Sweet corn – **a randomized block experiment**
  Correction factor $= 15,311.587$

Source	d.f.	Sum of squares	Mean square	Variance ratio	P
Density	4	380.804	95.201	36.124	<0.001
Block	4	64.491	16.123	6.118	<0.01
Residual	16	42.166	2.635		
TOTAL	24	487.461			

s.e.d.m. $= \sqrt{2 \times 2.635/5} = 1.03$
$t\,(P = 0.05)$ for 16 d.f. $= 2.12$

Therefore any means differing by more than $1.03 \times 2.12 = 2.18$ are significantly different.

	$20/m^2$		25		30		35		40
Means	**17.58**	<	**23.64**	<	**27.18**	=	**28.20**	=	**27.14**

2 Potatoes – **a fully randomized experiment**
Correction factor $= 1206.702$

Source	d.f.	Sum of squares	Mean square	Variance ratio	P
Species	3	372.625	124.208	29.394	<0.001
Residual	36	152.122	4.226		
TOTAL	39	524.747			

s.e.d.m. $= \sqrt{2 \times 4.226/10} = 0.92$
$t\,(P = 0.05)$ for 36 d.f. $= 2.04$

Therefore any means differing from the mean of B by more than $0.92 \times 2.04 = 1.88$ are significantly different.

	A	B	C	D
Means	1.30	5.11	5.65	9.91

Mean A is significantly lower and mean D significantly higher than the mean of B. However, in the test D–B, we are not testing two means adjacent in order of magnitude (5.65 lies in between). This increases the probability of the difference arising by chance and a multiple range test (see much later in Chapter 16) should really be used.

3 Lettuces – **a randomized block experiment**
Correction factor $= 360.622$

Source	d.f.	Sum of squares	Mean square	Variance ratio	P
Variety	5	1.737	0.347	115.667	<0.001
Block (= student)	7	0.056	0.008	2.667	ns
Residual	35	0.094	0.003		
TOTAL	47	1.887			

It is clear that students (= blocks, i.e. also position) **have had no significant effect on lettuce yields**, whereas seed packets have had a highly significant effect.

4 *Zinnia* plants – **a fully randomized experiment**

Correction factor = 183,384.14

Source	d.f.	Sum of squares	Mean square	Variance ratio	P
Treatment	2	406.157*	203.079	2.004	ns
Residual	36	3647.700	101.325		
TOTAL	27*	4053.857			

An LSD test is not necessary – **neither growth promoter significantly accelerates flowering** at least with this amount of replication! The means are 75.8 days for *P*-1049 and 80.7 for *P*-2711, compared with 85.3 days with no promoter.

*Effect of missing data for P-1049. Total plots are 28, giving 27 degrees of freedom. The added squares need to be calculated separately for each treatment as there will be a different divisor:

$$\text{Sum of squares for treatments} = \frac{853^2}{10} + \frac{606^2}{8} + \frac{807^2}{10} - \text{Correction factor}$$

Chapter 13 – Viticulture experiment

Correction factor = 2270.898

Source	d.f.	Sum of squares	Mean square	Variance ratio	P
Variety	3	178.871	59.624	34.27	<0.001
Training	1	11.057	11.057	6.35	<0.05
Interaction	3	13.560	4.520	2.60	ns
Treatments	7	203.488	29.070	16.71	<0.001
Blocks	4	14.247	3.562	2.05	ns
Residual	28	48.710	1.740		
TOTAL	39	266.445			

The interaction is NOT significant. All four varieties therefore respond similarly to the two training systems.

Chapter 14 – Plant species with fungicides and soils

Correction factor $= 264,180.17$

Source	d.f.	Sum of squares	Mean square	Variance ration	P
Species	2	9,900.11	4,950.06	52.60	<0.001
Soils	2	16,436.11	8,218.06	87.33	<0.001
Fungicide	1	1,932.02	1,932.02	20.53	<0.001
Species × Soils	4	658.44	164.61	1.75	ns
Species × Fungicides	2	194.03	97.02	1.03	ns
Soils × Fungicides	2	1,851.14	925.57	9.84	<0.001
Species × Soils × Fungicides	4	1,069.65	267.41	2.84	<0.05
Treatments	17	32,041.50	1,894.79	20.03	<0.001
Replicates	2	356.77	178.39	1.90	ns
Residual	34	3,199.40	94.10		
TOTAL	53	35,597.67			

The numbers already calculated for you when you were presented with the original data are shown in italics.

The ONE significant two-factor interaction is that between *Soils* and *Fungicides*.

Chapter 15 – Potato experiment

Correction factor $= 98,790.125$

Source of variation	Degrees of freedom	Sum of squares	Mean square	Variance ratio	P
Main plots					
Ridging	1	840.500	840.500	24.566	<0.01
Replicates	7	1017.875	145.411	4.250	<0.05
Main plot residual	7	239.500	34.214		
Main plot total	**15**	**2097.875**			
Sub-plots					
Tuber size	1	1800.000	1800.000	69.25	<0.001
Ridging × Tuber size	1	36.125	36.125	1.39	ns
Sub-plot residual	14	363.875	25.991		
Sub-plot total	**16**	**2200.000**			
TOTAL	31	4297.875			

The interaction (*Ridging* × *Tuber size*) is far from significant.

Chapter 16

1 The analysis of variance backwards – the lentil watering regime experiment.

For the solution to this exercise, I'll go into rather more detail than usual, as the route to the solution and the order of the steps involved is as much part of the solution as the final analysis table.

If you've succeeded with the exercise, you can skip the rest and go straight to the table to check your solution!

Step 1. Convert the table of means to a table of totals and supertotals, based on each figure in the original table being the mean of four replicates. Give each figure the appropriate subscript, showing how many plots contribute to the total:

	Water regimes					
	W1	W2	W3	W4	W5	
Planting dates						
D1	92.0_4	88.0_4	103.6_4	116.8_4	112.8_4	513.2_{20}
D2	86.4_4	133.6_4	122.0_4	128.8_4	82.8_4	553.6_{20}
D3	52.8_4	80.0_4	76.0_4	90.0_4	80.0_4	378.8_{20}
	231.2_{12}	301.6_{12}	301.6_{12}	335.6_{12}	275.6_{12}	$\mathbf{1445.6_{60}}$

Step 2. Calculate the correction factor: $1445.6^2/60 = 34,829.321$

Step 3. Compose the lead line: 3 planting dates $\times$ 5 watering regimes ($= 15$ treatments) $\times$ 4 replicates $= 60$ plots and from this the skeleton analysis of variance table:

Correction factor $= 34,829.321$

Source	d.f.	Sum of squares	Mean square	Variance ratio	P
Dates	2				
Waterings	4				
Interaction	8				
Treatments	14				
Replicates	3				
Residual	42				
TOTAL	59				

Step 4. This is where we start going backwards, by beginning with **PHASE 3.** Remember that Phase 3 in a factorial experiment uses

nothing else but the **supertotals** shown in italics at the ends of the columns and rows in the table of totals above. So we can do the **SqADS** in Phase 3 for the main effects of Dates and Waterings.

Step 5. **PHASE 2.** Again, remember Phase 2 utilizes only the replicate and individual treatment totals. We have the latter, so we can **SqADS** for treatments. By subtraction, we can then find the sum of squares for the interaction, the last item in the Phase 3 part of the analysis. We have to leave the replicate line blank. So far this gives us the following table (the numbers against each source of variation give the suggested order of calculation):

Correction factor = 34,829.321

Source	d.f.	Sum of squares	Mean square	Variance ratio	P
① Dates	2	837.511			
② Waterings	4	500.785			
④ Interaction	8	435.903			
③ Treatments	14	1774.199			
Replicates	3				
Residual	42				
TOTAL	59				

Step 6. **THE RESIDUAL SUM OF SQUARES.** Normally, we find the residual sum of squares by subtraction, but we cannot do this without the replicate sum of squares. So we have to work out the residual backwards from the LSD. If you remember, the LSD is:

$$t_{(P=0.05\text{for the 42 residual d.f.})} \times \text{s.e.d.m.}$$

where s.e.d.m. $= \sqrt{(2 \times \text{residual mean square})/n}$

2.021 is the nearest t in tables, and means are of 4 plots. Therefore the LSD, which is given as $5.7 = 2.021 \times \sqrt{(2 \times \text{residual mean square})/4}$. If we square everything, the equation becomes: $32.49 = 4.084 \times 2 \times \text{RMS}/4$. Now we have the RMS as the one unknown, which we can solve as:

$$2 \times \text{RMS} = \frac{32.49}{4.084} \times 4 = 31.822 \quad \text{so RMS} = 15.911$$

This is then calculation ⑤ in the table, and multiplying it by the residual d.f. (42) gets us backwards to calculation ⑥, the residual sum of squares. The rest of the analysis table can then be completed:

Correction factor = 34,829.321

Source	d.f.	Sum of squares	Mean square	Variance ratio	P
① Dates	2	837. 511	418.756	26.319	<0.001
② Waterings	4	500.785	125.196	7.869	<0.001
④ Interaction	8	435.903	54.488	3.425	<0.01
③ Treatments	14	1774.199	126.729	7.965	<0.001
Replicates	3	cannot be calculated			
Residual	42	⑥ 668.262	⑤ 15.911		
TOTAL	59	cannot be calculated			

I have done a full multiple range test on the data, and (arranging the means in order of magnitude of the main date and watering means) the following table of means and significant differences (as shown by the a,b,c ... notation) results:

	Water regimes				
	W4	W2	W3	W5	W1
Planting dates					
D2	32.2	33.4	30.5	20.7	21.6
	a	a	ab	cdef	bcdef
D1	29.2	22.0	25.9	28.2	23.0
	abc	bcde	abcd	abcd	bcd
D3	22.5	20.0	19.0	20.0	13.2
	bcd	cdef	ef	cdef	g

It is hard to find any general points to make from this table, other than that watering regimes do not significantly improve the nodulation of lentils planted on D1 in comparison with the unirrigated control (W5), and that high irrigation at either the vegetative (W3) or reproductive (W2 or W4) phase improves the nodulation of D2 plants over the control. The results for D3 plants are especially hard to interpret. High irrigation throughout (W4) givers superior nodulation to high irrigation only in the vegetative stage (W3), but the very low nodulation in W1 (low irrigation throughout) compared with the unirrigated control seems a strange phenomenon.

2 The viticulture experiment in Chapter 13.
   (a) As there is no interaction, there cannot be evidence that different
       varieties do better under different training systems.
   (b) For (b) it is worth having available the overall means of the two
       factors:

   Traditional   7.01.   High-wire   8.06   *F-test shows these means differ significantly.*

   For variety, LSD $= 2.021 \times \sqrt{(2 \times 1.74)/10} = 1.19$, so separation of means is:

   Huchselr. 9.69   Pinot B. 9.16   Semillon 6.98   Müller-Th. 4.32
        a   a   b   c

   With no interaction, the significant overall effect of training systems
means that the superiority of the high-wire system will increase the
yield of both varieties. But this could be cancelled out by the lower yield
of Pinot Blanc than Huchselrebe. The two means to test are 8.37 (the
5 plots of Trad. Huchs.) and 9.39 (the 5 plots of High W. Pinot). The
difference is 1.02 more on High W. Pinot. The appropriate LSD is now
that for means of 5, not 10, values. We therefore need to replace the 10
in the equation above for the LSD with 5 to get 1.69. We cannot therefore
say the 1.10 tonnes/ha increase from switching is real.

3 The fungicidal seed dressing experiment in Chapter 14.
   You were asked to interpret the *Soils × Fungicides* interaction represented by the following table of 6 means, each the mean of
9 plots.
   The means are as follows (again arranged in order of magnitude of
the main effects of the two factors):

	Sand	Silt loam	Clay
With Fungicide	86.88a	81.00ab	59.89c
Untreated	85.78a	74.44b	31.67d

   The LSD is $2.04 \times \sqrt{(2 \times 94.10)/9} = 9.33$. There are no differences
between nonadjacent means that require a multiple range test, and the
means separate as shown by the letters on the table above.
   The important horticultural feature of the interaction is that fungicide increases germination (compared with the untreated) significantly
ONLY in clay soil. This is because the increased moisture-holding capac-
ity of clay makes seed-rotting fungi more prevalent, i.e. there is not really
a problem for the fungicide to control in lighter soils! Another less important element in the interaction is that, whereas per cent germination is
significantly better in loam than sand with untreated seeds, the effect is
not significant where fungicide has been used.

4 The Chapter 15 split-plot potato experiment

As the interaction is not significant, and both factors are only at two levels, the F-tests in the analysis of variance are all that is needed to show that ridging yields significantly more tubers per plot (mean 60.69) than planting on the flat (mean 50.44) and that significantly more tubers per plot (mean 63.06) are produced by sowing large than small tubers (mean 48.06).

Chapter 17

1 Regression of weight (y) of bean seeds on length (x).

Regression coefficient (b) = 0.7397
Intercept at $x = 0$ (a) = -0.4890 (note the negative intercept; a bean of zero length weighs less than nothing!)

Therefore regression equation is Weight in g = $-0.4890 + (0.7397 \times$ length in cm). To draw the graph we use this equation to calculate 3 values of y from 3 values of x. The first is easy, for mean x (2.04) the value of y will be mean y (1.02).

The other two x values are up to you, but I usually use two fairly extreme values within the range of the data; here I pick x as 1.6 and 2.4:

$$\text{For } x = 1.6, y = -0.4890 + (0.7397 \times 1.6) = 0.695$$
$$\text{For } x = 2.04, y = -0.4890 + (0.7397 \times 2.4) = 1.286$$

Figure A3.1 is the graph you should have drawn.

To test the significance of the slope (b), we have to carry out an analysis of variance:

Source	d.f.	Sum squares	Mean square	Variance ratio	P
Regression	1	0.3196	0.3198	33.313	<0.001
Deviations	8	0.0764	0.0096		
TOTAL	9	0.3960			

Thus the slope of the regression is highly significant.

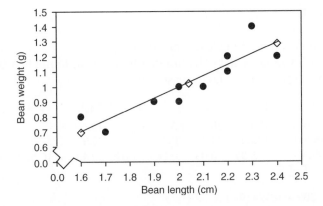

Fig. A3.1 Regression line of bean weight on bean length for "spare-time activity" 1 of Chapter 17. Closed circles, data points; open diamonds, points calculated in order to draw the regression line.

The regression has accounted for 0.3196 out of 0.3980 of the total variation (as sum of squares) of seed weight giving a coefficient of variation (r^2) of 0.8030.

2 Predicting yield of strawberry plants from number of flower heads.

$$\text{Regression coefficient } (b) = 0.0498$$
$$\text{Intercept at } x = 0 \; (a) = 0.4637$$

Therefore regression equation is Yield (kg) $= 0.4637 + (0.0498 \times$ number of flower heads)

To answer the second question, we have to carry out an analysis of variance:

Source	d.f.	Sum squares	Mean square	Variance ratio	P
Regression	1	0.3667	0.3667	18.592	<0.001
Deviations	13	0.2564	0.0197		
TOTAL	14	0.6231			

Thus the regression has accounted for 0.3667 out of 0.6231 of the total variation (as sum of squares) of yield from the mean yield. This is only 58.9%. Therefore, the relationship between yield and number of flower heads **is not** adequate to substitute a count of flower heads for a measurement of yield, given that 80% accuracy was required.

Chapter 18

1 Inspection and washing leaves for counting aphids.

Expected values will be the number in an instar found by washing $\times$ total by inspection $\div$ total by washing:

Instar	1	2	3	4	Adult	
Observed	105	96	83	62	43	
Expected	133.2*	94.4	70.4	52.1	38.9	
χ^2 (4d.f.)	5.99	0.03	2.26	1.88	0.43	$= 10.56$ ($P < 0.05$)

$$= \sum \frac{(O-E)^2}{E}$$

* i.e. $389 \times 161/470$

The observed data therefore deviate significantly from the expected. The largest χ^2 value is for the first instar, therefore run the test again with the first instar omitted. Expected values will now reflect the smaller inspection and washing totals 284 and 309 respectively.

Instar	2	3	4	Adult	
Observed	96	83	62	43	
Expected	104.8*	78.1	57.9	43.2	
χ^2 (3 d.f.)	0.74	0.31	0.29	0.00	$= 1.34$ (ns)

$$= \sum \frac{(O-E)^2}{E}$$

* i.e. $284 \times 114/309$

The observed and expected values now fit well. Clearly, inspection underestimates the proportion of aphids in the first instar (probably because some of these very small aphids are missed by eye).

2 Movement of snails on a brick wall

This analysis for χ^2 with heterogeneity involves numbers less than 10, so Yates' correction for continuity should be applied:

Trial	Upwards	Downwards	Expected value	χ^2 (1 d.f.)
1	7	3	5	0.90*
2	5	4	4.5	0.00
3	4	4	4	0.00
4	6	4	5	0.10
5	5	5	5	0.00
6	7	2	4.5	1.78
			χ^2(6 d.f.)	2.78
Overall	**34**	**22**	**28**	χ^2 (1 d.f.) **2.16**

*Example calculation for 7 *versus* 3, expected value $5 = 2 \times \frac{(2-0.5)^2}{5} = 0.90$

Analysis table:

	d.f.	χ^2	P
Overall	1	2.16	ns
Heterogeneity	5	0.62	ns
Total	6	2.78	

Neither the overall or heterogeneity χ^2 values are anywhere near significant. In spite of the apparent preference for snails to move upwards, the data are fully consistent with the hypothesis that movement is equal in both directions, and there seems no significant variation in this result between the trials. The number of snails is probably too small for this to be a convincing experiment.

3 Survival/mortality of a fly leaf miner

These data fit a 3 by 4 contingency table, which will therefore have $2 \times 3 = 6$ d.f. The expected values (based on the column and row end totals as well as the grand total, see page 290) for each cell are shown in italics and in brackets:

	Top	Middle	Bottom	Total
Healthy larva	9 *(14.62)*	41 *(35.02)*	18 *(18.36)*	68
Healthy pupa	11 *(8.60)*	21 *(20.60)*	6 *(10.80)*	40
Parasitoid pupa	8 *(10.97)*	19 *(26.27)*	24 *(13.77)*	51
Eaten by bird	15 *(8.82)*	20 *(21.12)*	6 *(11.07)*	41
Total	*43*	*103*	*54*	**200**

With two expected values below 10, it is appropriate to apply the Yates' correction for continuity in calculating the χ^2 values for each cell (added in bold and in square brackets):

	Top	Middle	Bottom
Healthy larva	9 *(14.62)* **[1.79]**	41 *(35.02)* **[0.86]**	18 *(18.36)* **[0.00]**
Healthy pupa	11 *(8.60)* **[0.42]**	21 *(20.60)* **[0.18]**	6 *(10.80)* **[1.71]**
Parasitoid pupa	8 *(10.97)* **[0.56]**	19 *(26.27)* **[1.74]**	24 *(13.77)* **[6.88]**
Eaten by bird	15 *(8.82)* **[3.66]**	20 *(21.12)* **[0.02]**	6 *(11.07)* **[1.89]**

The total of these χ^2 values is 19.71 for 6 d.f. ($P < 0.01$). There are clearly some significant associations between the frequencies of the fate of larvae and particular tree strata. Inspection of the table suggests the largest χ^2 values associate bird predation with the top of trees, and parasitization with the bottom. The results presumably reflect that birds and parasitoids search particularly in different tree strata.

The Clues (see "Spare-time activity" to Chapter 16, page 236)

Here are your two clues:

1 The figures in the table are MEANS of 4 replicates. So, when multiplied by four, they convert to the treatment TOTALS with a "4" subscript.
2 Since you cannot calculate the total sum of squares without all the original data, you will not be able to find the residual sum of squares by subtraction. However, you are given the LSD, and the formula for the LSD (page 215) includes the residual mean square as one of the components.

Appendix 4
Bibliography

Introduction

Just to repeat what I said in Chapter 1. This bibliography is no more than a list of a few books I have found particularly helpful over the years. The large number of different statistics books that have been published inevitably means that the vast majority are therefore not mentioned because I have no personal experience of using them – and it's virtually impossible to judge a statistics book in a library or bookshop. There you are alas more likely to find the books I haven't mentioned than the ones I have! Because my interest in statistics started a long time ago, this applies increasingly the more recently a book has been published! Even the books I mention will often have had more recent editions. However, even the oldest book I mention is still available secondhand on the internet.

Campbell, R.C. (1974). *Statistics for Biologists*, 2nd edition. Cambridge University Press: Cambridge, 385 pp.
This is the book I reach for to carry out nonparametric tests. I find straight-foward worked examples to follow, and the book also contains the tables for the related significance tests.

Goulden, C.H. (1952). *Methods of Statistical Analysis*, 2nd edition. John Wiley: New York, 467 pp.
This is the first statistics text I treated myself to, and its battered state shows how often I have consulted it. The worked examples are very easy to follow, and so I have used it mainly to expand my experience of statistical procedures, e.g. for calculating missing values in the analysis of variance, and multiple and nonlinear regression. It would be a good book for seeing if someone else's explanation will help where you find difficulty in my book.

Mead, R., Curnow, R.N. and Hasted, A.M. (1993). *Statistical Methods in Agriculture and Experimental Biology*, 2nd edition. Chapman and Hall: London, 415 pp.

Has good and well laid out examples and an excellent chapter on "Choosing the most appropriate experimental design." Otherwise the book is rather difficult for beginners in the reliance it places on statistical notation in describing procedures.

Schefler, W.C. (1983). *Statistics for Health Professionals.* **Addison-Wesley: Reading, Massachusetts, 302 pp.**

I have only recently come across this book, and I think you would find this one particularly helpful as a source of explanations of statistical theory by another author. It is perhaps also an excellent "next step" from my book to more advanced texts. I can particularly recommend the section on chi-square.

Snedecor G.W. and Cochran, W.G. (1967). *Statistical Methods,* **6th edition. Iowa State University Press: Ames, Iowa, 593 pp.**

This is one of my standard reference books for calculation formulae as well as descriptions and worked examples of more advanced statistical techniques.

Steel, R.G.D. and Torrie, J.H. (1960). *Principles and Procedures of Statistics with Special Reference to the Biological Sciences.* **McGraw-Hill: New York, 481 pp.**

This is my favorite standard reference work for reading up the theory and reminding myself of computational procedures. I find it particularly easy to locate what I want without having to browse large sections of text. It is, however, rather algebraic in approach.

Index